Climatology

Climatology

Dominic Pratt

Larsen & Keller
www.larsen-keller.com

Climatology
Dominic Pratt
ISBN: 978-1-64172-665-8 (Hardback)

🖰 Larsen & Keller

Published by Larsen and Keller Education,
5 Penn Plaza,
19th Floor,
New York, NY 10001, USA

Cataloging-in-Publication Data

Climatology / Dominic Pratt.
    p. cm.
Includes bibliographical references and index.
ISBN 978-1-64172-665-8
1. Climatology. 2. Atmospheric science. 3. Climatic changes. I. Pratt, Dominic.
QC861.3 .C55 2022
551.6--dc23

For more information regarding Larsen and Keller Education and its products, please visit the publisher's website www.larsen-keller.com

TABLE OF CONTENTS

PREFACE

The purpose of this book is to help students understand the fundamental concepts of this discipline. It is designed to motivate students to learn and prosper. I am grateful for the support of my colleagues. I would also like to acknowledge the encouragement of my family.

The scientific study of climate falls under the domain of climatology. It is a branch of atmospheric science. It is closely related to the fields of oceanography and biogeochemistry. Climatology primarily deals with the analysis and modeling of the physical laws that determine the climate. The climate models are used for numerous purposes such as studying the dynamics of the weather and creating future climate projections. Climatology is broadly divided into three subcategories, on the basis of the purpose and complexity of research. These are scientific climatology, descriptive climatology and applied climatology. Some of the other sub-fields of climatology are paleoclimatology, historical climatology and boundary-layer climatology. This textbook is a valuable compilation of topics, ranging from the basic to the most complex theories and principles in the field of climatology. Some of the diverse topics covered herein address the varied branches that fall under this category. This book will provide comprehensive knowledge to the readers.

A foreword for all the chapters is provided below:

Chapter – What is Climate?

The long-term average of weather is termed as climate. The major factors that affect climate are ocean-currents, distance from the sea, direction of prevailing winds, shape of the land and distance from the equator. This chapter briefly introduces the climate, its classification according to the Köppen climate classification system and the factors that affect it.

Chapter – Climatology: Introduction

The scientific study of the weather conditions averaged over a period of time is referred to as climatology. It is a branch of atmospheric sciences and a sub-field of physical geography. Some of the elements of weather which are studied within this field are temperature, precipitation and wind. This chapter has been carefully written to provide an easy understanding of the varied facets of climatology.

Chapter – Sub-divisions of Climatology

Climatology can be sub-divided into numerous fields such as paleoclimatology, paleotempestology, tropical cyclone rainfall climatology, bioclimatology, historical climatology, urban climatology, dendroclimatology, synoptic climatology, hydroclimatology, etc. The topics elaborated in this chapter will help in gaining a better perspective about these sub-fields of climatology.

Chapter – Climate Change

Climate change refers to the changes in the climate system of the Earth that result in new weather patterns that remain the same for a long period of time. Some of the major areas of study within this field are global warming and greenhouse effect. The topics elaborated in this chapter will help in developing a thorough understanding of climate change, its effects and the ways to mitigate them.

Chapter – Models of Climate

Climate models are systems of differential equations which are used to simulate the interactions of the major drivers of climate such as land surface, oceans, ice and the atmosphere. A few of the important models are atmospheric model and general circulation model. This chapter closely examines these models of climate to provide an extensive understanding of the subject.

Dominic Pratt

What is Climate?

The long-term average of weather is termed as climate. The major factors that affect climate are ocean-currents, distance from the sea, direction of prevailing winds, shape of the land and distance from the equator. This chapter briefly introduces the climate, its classification according to the Köppen climate classification system and the factors that affect it.

Climate is the long-term pattern of weather in a particular area. Weather can change from hour-to-hour, day-to-day, month-to-month or even year-to-year. A region's weather patterns, usually tracked for at least 30 years, are considered its climate.

Climate System

Different parts of the world have different climates. Some parts of the world are hot and rainy nearly every day. They have a tropical wet climate. Others are cold and snow-covered most of the year. They have a polarclimate. Between the icy poles and the steamy tropics are many other climates that contribute to Earth's biodiversity and geologic heritage.

Climate is determined by a region's climate system. A climate system has five major components: the atmosphere, the hydrosphere, the cryosphere, the land surface, and the biosphere.

The atmosphere is the most variable part of the climate system. The composition and movement of gases surrounding the Earth can change radically, influenced by natural and man-made factors.

Changes to the hydrosphere, which include variations in temperature and salinity, occur at much slower rates than changes to the atmosphere.

The cryosphere is another generally consistent part of the climate system. Ice sheets and glaciers reflect sunlight, and the thermal conductivity of ice and permafrost profoundly influences temperature. The cryosphere also helps regulate thermohaline circulation. This "ocean conveyor belt" has an enormous influence on marine ecosystems and biodiversity.

Topography and vegetation influence climate by helping determine how the Sun's energy is used on Earth. The abundance of plants and the type of land cover (such as soil, sand, or asphalt) impacts evaporation and ambient temperature.

The biosphere, the sum total of living things on Earth, profoundly influences climate. Through photosynthesis, plants help regulate the flow of greenhouse gases in the atmosphere. Forests and oceans serve as "carbon sinks" that have a cooling impact on climate. Living organisms alter the landscape, through both natural growth and created structures such as burrows, dams, and mounds. These altered landscapes can influence weather patterns such as wind, erosion, and even temperature.

Climate Features

The most familiar features of a region's climate are probably average temperature and precipitation. Changes in day-to-day, day-to-night, and seasonal variations also help determine specific climates. For example, San Francisco, California, and Beijing, China, have similar yearly temperatures and precipitation. However, the daily and seasonal changes make San Francisco and Beijing very different. San Francisco's winters are not much cooler than its summers, while Beijing is hot in summer and cold in winter. San Francisco's summers are dry and its winters are wet. Wet and dry seasons are reversed in Beijing—it has rainy summers and dry winters.

Climate features also include windiness, humidity, cloud cover, atmospheric pressure, and fogginess. Latitude plays a huge factor in determining climate. Landscape can also help define regional climate. A region's elevation, proximity to the ocean or freshwater, and land-use patterns can all impact climate.

All climates are the product of many factors, including latitude, elevation, topography, distance from the ocean, and location on a continent. The rainy, tropical climate of West Africa, for example, is influenced by the region's location near the Equator (latitude) and its position on the western side of the continent. The area receives direct sunlight year-round, and sits at an area called the intertropical convergence zone, where moist trade winds meet. As a result, the region's climate is warm and rainy.

Microclimates

Of course, no climate is uniform. Small variations, called microclimates, exist in every climate region. Microclimates are largely influenced by topographic features such as lakes, vegetation, and cities. In large urban areas, for example, streets and buildings absorb heat from the Sun, raising the average temperature of the city higher than average temperatures of more open areas nearby. This is known as the "urban heat island effect."

Large bodies of water, such as the Great Lakes in the United States and Canada, can also have microclimates. Cities on the southern side of Lake Ontario, for example, are cloudier and receive much more snow than cities on the northern shore. This "lake effect" is a result of cold winds blowing across warmer lake water.

Factors affecting Climate

There are many different factors that affect climate around the world. It is the varying influence of these factors that lead to different parts of the Earth experiencing differing climates. The most important natural factors are:

- Distance from the sea,

- Ocean currents,

- Direction of prevailing winds,

- Shape of the land (known as 'relief' or 'topography'),

- Distance from the equator,

- The El Niño phenomenon.

It is now widely accepted that human activity is also affecting climate, and that the impact is not the same everywhere. For example, changes appear to be happening faster near the poles than in many other places.

Distance from the Sea (Continentality)

The sea affects the climate of a place. Coastal areas are cooler and wetter than inland areas. Clouds form when warm air from inland areas meets cool air from the sea. The centre of continents are subject to a large range of temperatures. In the summer, temperatures can be very hot and dry as moisture from the sea evaporates before it reaches the centre of the land mass.

Ocean Currents

Ocean currents can increase or reduce temperatures. The diagram below shows the ocean currents of the world. The main ocean current that affects the UK is the Gulf Stream.

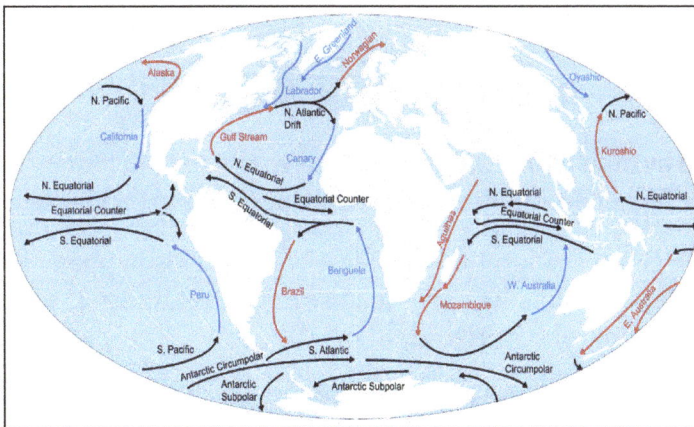

The Gulf Stream is a warm ocean current in the North Atlantic flowing from the Gulf of Mexico, northeast along the U.S coast, and from there to the British Isles.

The Gulf of Mexico has higher air temperatures than Britain as it is closer to the equator. This means that the air coming from the Gulf of Mexico to Britain is also warm. However, the air is also quite moist as it travels over the Atlantic ocean. This is one reason why Britain often receives wet weather.

The Gulf Stream keeps the west coast of Europe free from ice in the winter and, in the summer, warmer than other places of a similar latitude.

Direction of Prevailing Winds

Winds that blow from the sea often bring rain to the coast and dry weather to inland areas. Winds that blow to Britain from warm inland areas such as Africa will be warm and dry. Winds that blow to Britain from inland areas such as central Europe will be cold and dry in winter. Britain's prevailing (i.e. most frequently experienced) winds come from a south westerly direction over the Atlantic. These winds are cool in the summer, mild in the winter and tend to bring wet weather.

The Shape of the Land

Climate can be affected by mountains. Mountains receive more rainfall than low lying areas because as air is forced over the higher ground it cools, causing moist air to condense and fall out as rainfall.

The higher the place is above sea level the colder it will be. This happens because as altitude increases, air becomes thinner and is less able to absorb and retain heat. That is why you may see snow on the top of mountains all year round.

Distance from the Equator

The distance from the equator affects the climate of a place. At the poles, energy from the sun reaches the Earth's surface at lower angles and passes through a thicker layer of atmosphere than at the equator. This means the climate is cooler further from the Equator. The poles also experience the greatest difference between summer and winter day lengths: in the summer there is a period when the sun does not set at the poles; conversely the poles also experience a period of total darkness during winter. In contrast, daylength varies little at the equator.

El Niño

El Niño, which affects wind and rainfall patterns, has been blamed for droughts and floods in countries around the Pacific Rim. *El Niño* refers to the irregular warming of surface water in the Pacific. The warmer water pumps energy and moisture into the atmosphere, altering global wind and rainfall patterns. The phenomenon has caused tornadoes in Florida, smog in Indonesia, and forest fires in Brazil.

El Niño is Spanish for 'the Boy Child' because it comes about the time of the celebration of the birth of the Christ Child. The cold counterpart to *El Niño* is known as *La Niña*, Spanish for 'the girl child', and it also brings with it weather extremes.

Human Influence

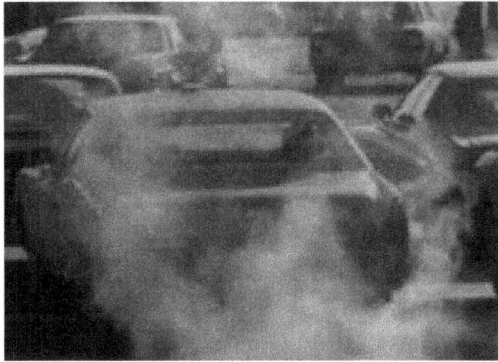

The factors above affect the climate naturally. However, we cannot forget the influence of humans on our climate. Early on in human history our effect on the climate would have been quite small. However, as populations increased and trees were cut down in large numbers, so our influence on the climate increased. Trees take in carbon dioxide and produce oxygen. A reduction in trees will therefore have increased the amount of carbon dioxide in the atmosphere.

The Industrial Revolution, starting at the end of the 19th Century, has had a huge effect on climate. The invention of the motor engine and the increased burning of fossil fuels have increased the amount of carbon dioxide (a greenhouse gas - more on that later) in the atmosphere. The number of trees being cut down has also increased, reducing the amount of carbon dioxide that is taken up by forests.

Urban Climate

Urban Climate is any set of climatic conditions that prevails in a large metropolitan area and that differs from the climate of its rural surroundings.

Urban climates are distinguished from those of less built-up areas by differences of air temperature, humidity, wind speed and direction, and amount of precipitation. These differences are attributable in large part to the altering of the natural terrain through the construction of artificial structures and surfaces. For example, tall buildings, paved streets, and parking lots affect wind flow, precipitation runoff, and the energy balance of a locale.

Also characteristic of the atmosphere over urban centres are substantially higher concentrations of pollutants such as carbon monoxide, the oxides of sulfur and nitrogen, hydrocarbons, oxidants, and particulate matter. Foreign matter of this kind is introduced into the air by industrial processes (e.g., chemical discharges by oil refineries), fuel combustion (for the operation of motor vehicles and for the heating of offices and factories), and the burning of solid wastes. Urban pollution concentrations depend on the magnitude of local emissions sources and the prevailing meteorological ventilation of the area—i.e., the height of the atmospheric layer through which the pollutants are being mixed and the average wind speed through that layer. Heavy concentrations of air pollutants have considerable impact on temperature, visibility, and precipitation in and around cities. Moreover, there occasionally arise weather conditions that allow the accumulation of pollutants over an urban area for several days. Such conditions, termed temperature inversions (increasing air temperature with increasing altitude), strongly inhibit atmospheric mixing and can cause acute distress in the population and even, under extremely severe conditions, loss of life. Atmospheric inversion caused an air-pollution disaster in London in December 1952 in which about 3,500 persons died from respiratory diseases.

The centre of a city is warmer than are outlying areas. Daily minimum temperature readings at related urban and rural sites frequently show that the urban site is 6° to 11° C (10° to 20° F) warmer than the rural site. Two primary processes influence the formation of this "heat island." During summer, urban masonry and asphalt absorb, store, and reradiate more solar energy per unit area than do the vegetation and soil typical of rural areas. Furthermore, less of this energy can be used for evaporation in urban areas, which characteristically exhibit greater precipitation runoff from streets and buildings. At night, radiative losses from urban building and street materials keep the city's air warmer than that of rural areas.

During winter the urban atmosphere is warmed slightly, but significantly, by energy from fuel combustion for home heating, power generation, industry, and transportation. Also contributing to the warmer urban atmosphere is the blanket of pollutants and water vapour that absorbs a portion of the thermal radiation emitted by the Earth's surface. Part of the absorbed radiation warms the surrounding air, a process that tends to stabilize the air over a city, which in turn increases the probability of higher pollutant concentrations.

The average relative humidity in cities is usually several percent lower than that of adjacent rural areas, primarily because of increased runoff of precipitation and the lack of evapotranspiration from vegetation in urban areas. Some moisture, however, is added to urban atmospheres by the many combustion sources.

The flow of wind through a city is characterized by mean speeds that are 20 to 30 percent lower than those of winds blowing across the adjacent countryside. This difference occurs as a result of the increased frictional drag on air flowing over built-up urban terrain, which is rougher than rural areas. Another difference between urban and rural wind flow is the convergence of low-level wind over a city (i.e., air tends to flow into a city from all directions). This is caused primarily by the horizontal thermal gradients of the urban heat island.

The amount of solar radiation received by cities is reduced by the blanket of particulates in the overlying atmosphere. The higher particulate concentrations in urban atmospheres reduce visibility by both scattering and absorbing light. In addition, some particles provide opportunities for the condensation of water vapour to form water droplets, the ingredients of fog.

A city also influences precipitation patterns in its vicinity. Such city-generated or city-modified weather factors as wind turbulence, thermal convection, and high concentrations of condensation nuclei might be expected to increase precipitation. Although appropriate continuous, quantitative measurements have not been made for a sufficient length of time, there is some data to suggest that the amount of precipitation over many large cities is about 5 to 10 percent greater than that over nearby rural areas, with the greatest increases occurring downwind of the city centre.

Urban Climatology

An urban climatologist studies the climate in and around cities. Urban areas are both affected by weather and climate, and exert an influence on the local scale weather and climate. The climate in and around cities and other built up areas is altered in part due to modifications humans make to the surface of the Earth during urbanization. The surface is typically rougher and often drier in cities, as naturally vegetated surfaces are replaced by buildings and paved streets. Buildings along streets form urban street "canyons" that cause the urban surface to take on a distinctly three-dimensional character. These changes affect the absorption of solar radiation, the surface temperature, evaporation rates, storage of heat and the turbulence and wind climates of cities and can drastically alter the conditions of the near-surface atmosphere. Human activities in cities also produce emissions of heat, water vapour and pollutants that directly impact the temperature, humidity, visibility and air quality in the atmosphere above cities. On slightly larger scales, urbanization can also lead to changes in precipitation above and downwind of urban areas. In fact, urbanization alters just about every element of climate and weather in the atmosphere above the city, and sometimes downwind of the city.

Although cities themselves form a very small fraction of the Earth's surface area, the world's population has become increasingly urbanized and is now affected by urban climates. Cities too, are important sites for greenhouse gas emissions because of the high energy demands by urban residents and activities. These emissions extend the (indirect) influence of cities on climate to much larger scales. Locally altered urban climates that have existed for many years may provide us with some insight into how to respond to large scale climate change and this makes the study of urban climates increasingly important.

The author (left photograph) preparing instruments for an urban climate study in Tokyo, Japan. The right-hand photograph shows the extended tower with the skyline of the Shinjuku area of Tokyo in the background.

Urban climatologists study a range of problems. Some urban climatologists specialize in understanding atmospheric processes, and therefore use knowledge and techniques from the fields of geography, meteorology, chemistry, physics and biology to study problems such as air pollution, turbulence, precipitation effects, urban heat islands, and representing urban areas in numerical models. Others study more applied problems such as: urban climate and the health of city inhabitants, the use of greenspace and vegetation to improve urban climates, urban climate and urban planning, or the use of remote sensing to observe urban climates.

Subnivean Climate

Subnivean climate is the environment of many hibernal animals, as it provides insulation and protection from predators. The subnivean climate is formed by three different types of snow metamorphosis: destructive metamorphosis, which begins when snow falls; constructive metamorphosis, the movement of water vapor to the surface of the snowpack; and melt metamorphosis, the melting/sublimation of snow to water vapor and its refreezing in the snowpack. These three types of metamorphosis transform individual snowflakes into ice crystals and create spaces under the snow where small animals can move.

Subnivean Fauna

Subnivean fauna includes small mammals such as mice, voles, shrews, and lemmings that must rely on winter snow cover for survival. These mammals move under the snow for protection from heat loss and some predators. In winter regions that do not have permafrost, the subnivean zone maintains a temperature of close to 32 °F (0 °C) regardless of the temperature above the snow cover, once the snow cover has reached a depth of six inches (15 cm) or more. The sinuous tunnels left by these small mammals can be seen from above when the snow melts to the final inch or so.

Some winter predators, such as foxes and large owls, can hear their prey through the snow and pounce from above. Ermine (stoats) can enter and hunt below the snowpack. Snowmobiles and ATVs can collapse the subnivean space. Skis and snow shoes are less likely to collapse subnivean space if the snowpack is deep enough.

Larger animals also use subnivean space. In the Arctic, ringed seals have closed spaces under the snow and above openings in the ice. In addition to resting and sleeping there, the female seals give birth to their pups on the ice. Female polar bears also den in snow caves to give birth to their young. Both types of dens are protected from exterior temperatures. Formation of these large spaces is from the animals' activity, not ground heat.

Subnivean Climate Formation

Deconstructive Metamorphosis

Deconstructive metamorphosis begins as the snow makes its way to the ground, often melting, refreezing, and settling. Water molecules become reordered, causing the snowflakes to become more spherical in appearance. These melting snowflakes fuse with others around them, becoming larger until all are uniform in size. While the snow is on the ground, the melting and joining of snow flakes reduces the height of snowpack by shrinking air spaces, causing the density and mechanical strength of the snowpack to increase. Freshly fallen snow with a density of 0.1 g/cm³ has very good insulating properties; however as time goes on, due to destructive metamorphism, the insulating property of the snowpack decreases, because the air spaces between snowflakes disappear. Snow that has been on the ground for a long period of time has an average density of 0.40 g/cm^3 and conducts heat well; however, once a base of 50 cm of snow with a density around 0.3 g/cm^3 has accumulated, temperatures under the snow remain relatively constant because the greater depth of snow compensates for its density. Destructive metamorphosis is a function of time, location, and weather. It occurs at a faster rate with higher temperatures, in the presence of water, under larger temperature gradients (e.g., warm days followed by cold nights), at lower elevations, and on slopes that receive large amounts of solar radiation. As time goes on, snow settles, compacting air spaces, a process expedited by the packing force of the wind.

Compaction of snow reduces the penetration of long- and short-wave radiation by reflecting more radiation off the snow. This limitation of light transmission through the snowpack decreases light availability under the snow. Only 3% of light can penetrate to a depth of 20 cm of snow when the density is 0.21 g/cm^3. At a depth of 40 cm, less than 0.2% of light is transmitted from the snow surface to ground below. This decrease in light transmission occurs up to the point at which critical compaction is reached. This occurs because the surface area of the ice crystal decreases and it causes less refraction and scattering of light. Once densities reach 0.5 g/cm^3, total surface area is reduced, which in turn reduces internal refraction and allows light to penetrate deeper into the snowpack.

Constructive Metamorphosis

Constructive metamorphosis is caused by the upward movement of water vapor within the snowpack. Warmer temperatures are found closer to the ground because it receives heat from the core of the earth. Snow has a low thermal conductivity, so this heat is retained, creating a temperature gradient between the air underneath the snowpack and the air above it. Warmer air holds more water vapor. Through the process of sublimation, the newly formed water vapor travels vertically by way of diffusion from a higher concentration (next to the ground) to a lower concentration (near the snowpack surface) by traveling through the air spaces between ice crystals. When the water vapor reaches the top of the snowpack, it is subjected to much colder air, causing it to condense and refreeze, forming ice crystals at the top of the snowpack that can be seen as the layer of crust on top of the snow.

Melt Metamorphism

Melt metamorphism is the deterioration of snow by melting. Melting can be stimulated by warmer ambient temperatures, rain, and fog. As snow melts, water is formed and the force of gravity pulls these molecules downward. En route to the ground, they refreeze, thickening in the middle stratum. During this refreezing process, energy is released in the form of latent heat. As more water comes down from the surface, it creates more heat and brings the entire snowpack column to near equal temperature. The firnification of the snow strengthens the snowpack, due to the bonding of grains of snow. Snow around trees and under canopies melts faster due to the reradiation of long-wave radiation. As snow gets older, particles of impurities (pine needles, soil, and leaves, for example) accrue within the snow. These darkened objects absorb more short-wave radiation, causing them to rise in temperature, also reflecting more long-wave radiation.

Köppen Climate Classification

The Köppen climate classification is one of the most widely used climate classification systems. It was first published by the German-Russian climatologist Wladimir Köppen in 1884, with several later modifications by Köppen, notably in 1918 and 1936. Later, the climatologist Rudolf Geiger introduced some changes to the classification system, which is thus sometimes called the Köppen–Geiger climate classification system.

The Köppen climate classification divides climates into five main climate groups, with each group being divided based on seasonal precipitation and temperature patterns. The five main groups are *A* (tropical), *B* (dry), *C* (temperate), *D* (continental), and *E* (polar). Each group and subgroup is represented by a letter. All climates are assigned a main group (the first letter). All climates except for those in the *E* group are assigned a seasonal precipitation subgroup (the second letter). For example, *Af* indicates a tropical rainforest climate. The system assigns a temperature subgroup for all groups other than those in the *A* group, indicated by the third letter for climates in *B*, *C*, and *D*, and the second letter for climates in *E*. For example, *Cfb* indicates an oceanic climate with warm summers as indicated by the ending *b*. Climates are classified based on specific criteria unique to each climate type.

As Köppen designed the system based on his experience as a botanist, his main climate groups are based on what types of vegetation grow in a given climate classification region. In addition to identifying climates, the system can be used to analyze ecosystem conditions and identify the main types of vegetation within climates. Due to its link with the plant life of a given region, the system is useful in predicting future changes in plant life within that region.

The Köppen climate classification system has been further modified, within the Trewartha climate classification system in the middle 1960s (revised in 1980). The Trewartha system sought to create a more refined middle latitude climate zone, which was one of the criticisms of the Köppen system (the C climate group was too broad).

The Köppen climate classification scheme divides climates into five main climate groups: *A* (tropical), *B* (dry), *C* (temperate), *D* (continental), and *E* (polar). The second letter indicates the seasonal precipitation type, while the third letter indicates the level of heat. Summers are defined as the 6-month period that is warmer either from April–September and/or October–March while winter is the 6-month period that is cooler.

Group A: Tropical (Megathermal) Climates

This type of climate has every month of the year with an average temperature of 18 °C (64.4 °F) or higher, with significant precipitation.

- *Af* = Tropical rainforest climate; average precipitation of at least 60 mm (2.4 in) in every month.

- *Am* = Tropical monsoon climate; driest month (which nearly always occurs at or soon after the "winter" solstice for that side of the equator) with precipitation less than 60 mm (2.4 in), but at least $100 - \left(\dfrac{Total\ Annual\ Precipitation\ (mm)}{25} \right)$.

- *Aw* or *As* = Tropical wet and dry or savanna climate; with the driest month having precipitation less than 60 mm (2.4 in) and less than $100 - \left(\dfrac{Total\ Annual\ Precipitation\ (mm)}{25} \right)$.

Group B: Dry (Arid and Semiarid) Climates

This type of climate is defined by little precipitation. Multiply the average annual temperature in Celsius by 20, then add:

- 280 if 70% or more of the total precipitation is in the spring and summer months (April–September in the Northern Hemisphere, or October–March in the Southern),

- 140 if 30%–70% of the total precipitation is received during the spring and summer,

- 0 if less than 30% of the total precipitation is received during the spring and summer.

If the annual precipitation is less than 50% of this threshold, the classification is BW (arid: desert climate); if it is in the range of 50%–100% of the threshold, the classification is BS (semi-arid: steppe climate).

A third letter can be included to indicate temperature. Originally, h signified low-latitude climate (average annual temperature above 18 °C (64.4 °F)) while k signified middle-latitude climate (average annual temperature below 18 °C), but the more common practice today, especially in the United States, is to use h to mean the coldest month has an average temperature above 0 °C (32 °F) (or −3 °C (27 °F)), with k denoting that at least one month's averages below 0 °C (or −3 °C (27 °F)). The n is used to denote a climate characterized by frequent fog.

- *BWh* = Hot desert climate

- *BWk* = Cold desert climate

- *BSh* = Hot semi-arid climate

- *BSk* = Cold semi-arid climate

Köppen climate classification scheme symbols description table		
1st	2nd	3rd
A (Tropical)	f (Rainforest)	
	m (Monsoon)	
	w (Savanna, Wet summer)	
	s (Savanna, Dry summer)	
B (Arid)	W (Desert)	
	S (Steppe)	
		h (Hot)
		k (Cold)

C (Temperate)	s (Dry summer)	
	w (Dry winter)	
	f (Without dry season)	
		a (Hot summer)
		b (Warm summer)
		c (Cold summer)
D (Continental)	s (Dry summer)	
	w (Dry winter)	
	f (Without dry season)	
		a (Hot summer)
		b (Warm summer)
		c (Cold summer)
		d (Very cold winter)
E (Polar)	T (Tundra)	
	F (Eternal winter (ice cap))	

Group C: Temperate (Mesothermal) Climates

This type of climate has the coldest month averaging between 0 °C (32 °F) (or –3 °C (27 °F)) and 18 °C (64.4 °F) and at least one month averaging above 10 °C (50 °F).

- *Cfa* = Humid subtropical climate; coldest month averaging above 0 °C (32 °F) (or –3 °C (27 °F)), at least one month's average temperature above 22 °C (71.6 °F), and at least four months averaging above 10 °C (50 °F). No significant precipitation difference between seasons (neither abovementioned set of conditions fulfilled). No dry months in the summer.

- *Cfb* = Temperate oceanic climate; coldest month averaging above 0 °C (32 °F) (or –3 °C (27 °F)), all months with average temperatures below 22 °C (71.6 °F), and at least four months averaging above 10 °C (50 °F). No significant precipitation difference between seasons (neither abovementioned set of conditions fulfilled).

- *Cfc* = Subpolar oceanic climate; coldest month averaging above 0 °C (32 °F) (or –3 °C (27 °F)) and 1–3 months averaging above 10 °C (50 °F). No significant precipitation difference between seasons (neither abovementioned set of conditions fulfilled).

- *Cwa* = Monsoon-influenced humid subtropical climate; coldest month averaging above 0 °C (32 °F) (or –3 °C (27 °F)), at least one month's average temperature above 22 °C (71.6 °F), and at least four months averaging above 10 °C (50 °F). At least ten times as much rain in the wettest month of summer as in the driest month of winter (alternative definition is 70% or more of average annual precipitation is received in the warmest six months).

- *Cwb* = Subtropical highland climate or Monsoon-influenced temperate oceanic climate; coldest month averaging above 0 °C (32 °F) (or −3 °C (27 °F)), all months with average temperatures below 22 °C (71.6 °F), and at least four months averaging above 10 °C (50 °F). At least ten times as much rain in the wettest month of summer as in the driest month of winter (an alternative definition is 70% or more of average annual precipitation received in the warmest six months).

- *Cwc* = Cold subtropical highland climate or Monsoon-influenced subpolar oceanic climate; coldest month averaging above 0 °C (32 °F) (or −3 °C (27 °F)) and 1–3 months averaging above 10 °C (50 °F). At least ten times as much rain in the wettest month of summer as in the driest month of winter (alternative definition is 70% or more of average annual precipitation is received in the warmest six months).

- *Csa* = Hot-summer Mediterranean climate; coldest month averaging above 0 °C (32 °F) (or −3 °C (27 °F)), at least one month's average temperature above 22 °C (71.6 °F), and at least four months averaging above 10 °C (50 °F). At least three times as much precipitation in the wettest month of winter as in the driest month of summer, and driest month of summer receives less than 30 mm (1.2 in).

- *Csb* = Warm-summer Mediterranean climate; coldest month averaging above 0 °C (32 °F) (or −3 °C (27 °F)), all months with average temperatures below 22 °C (71.6 °F), and at least four months averaging above 10 °C (50 °F). At least three times as much precipitation in the wettest month of winter as in the driest month of summer, and driest month of summer receives less than 30 mm (1.2 in).

- *Csc* = Cold-summer Mediterranean climate; coldest month averaging above 0 °C (32 °F) (or −3 °C (27 °F)) and 1–3 months averaging above 10 °C (50 °F). At least three times as much precipitation in the wettest month of winter as in the driest month of summer, and driest month of summer receives less than 30 mm (1.2 in).

Group D: Continental (Microthermal) Climates

This type of climate has at least one month averaging below 0 °C (32 °F) (or −3 °C (27 °F)) and at least one month averaging above 10 °C (50 °F).

- *Dfa* = Hot-summer humid continental climate; coldest month averaging below −0 °C (32 °F) (or −3 °C (27 °F)), at least one month's average temperature above 22 °C (71.6 °F), and at least four months averaging above 10 °C (50 °F). No significant precipitation difference between seasons (neither above mentioned set of conditions fulfilled).

- *Dfb* = Warm-summer humid continental climate; coldest month averaging below −0 °C (32 °F) (or −3 °C (27 °F)), all months with average temperatures below 22 °C (71.6 °F), and at least four months averaging above 10 °C (50 °F). No significant precipitation difference between seasons (neither above mentioned set of conditions fulfilled).

- *Dfc* = Subarctic climate; coldest month averaging below 0 °C (32 °F) (or −3 °C (27 °F)) and 1–3 months averaging above 10 °C (50 °F). No significant precipitation difference between seasons (neither above mentioned set of conditions fulfilled).

- *Dfd* = Extremely cold subarctic climate; coldest month averaging below −38 °C (−36.4 °F) and 1−3 months averaging above 10 °C (50 °F). No significant precipitation difference between seasons (neither above mentioned set of conditions fulfilled).

- *Dwa* = Monsoon-influenced hot-summer humid continental climate; coldest month averaging below 0 °C (32 °F) (or −3 °C (27 °F)), at least one month's average temperature above 22 °C (71.6 °F), and at least four months averaging above 10 °C (50 °F). At least ten times as much rain in the wettest month of summer as in the driest month of winter (alternative definition is 70% or more of average annual precipitation is received in the warmest six months).

- *Dwb* = Monsoon-influenced warm-summer humid continental climate; coldest month averaging below 0 °C (32 °F) (or −3 °C (27 °F)), all months with average temperatures below 22 °C (71.6 °F), and at least four months averaging above 10 °C (50 °F). At least ten times as much rain in the wettest month of summer as in the driest month of winter (alternative definition is 70% or more of average annual precipitation is received in the warmest six months).

- *Dwc* = Monsoon-influenced subarctic climate; coldest month averaging below 0 °C (32 °F) (or −3 °C (27 °F)) and 1−3 months averaging above 10 °C (50 °F). At least ten times as much rain in the wettest month of summer as in the driest month of winter (alternative definition is 70% or more of average annual precipitation is received in the warmest six months).

- *Dwd* = Monsoon-influenced extremely cold subarctic climate; coldest month averaging below −38 °C (−36.4 °F) and 1−3 months averaging above 10 °C (50 °F). At least ten times as much rain in the wettest month of summer as in the driest month of winter (alternative definition is 70% or more of average annual precipitation is received in the warmest six months).

- *Dsa* = Mediterranean-influenced hot-summer humid continental climate; coldest month averaging below 0 °C (32 °F) (or −3 °C (27 °F)), average temperature of the warmest month above 22 °C (71.6 °F) and at least four months averaging above 10 °C (50 °F). At least three times as much precipitation in the wettest month of winter as in the driest month of summer, and driest month of summer receives less than 30 mm (1.2 in).

- *Dsb* = Mediterranean-influenced warm-summer humid continental climate; coldest month averaging below 0 °C (32 °F) (or −3 °C (27 °F)), average temperature of the warmest month below 22 °C (71.6 °F) and at least four months averaging above 10 °C (50 °F). At least three times as much precipitation in the wettest month of winter as in the driest month of summer, and driest month of summer receives less than 30 mm (1.2 in).

- *Dsc* = Mediterranean-influenced subarctic climate; coldest month averaging below 0 °C (32 °F) (or −3 °C (27 °F)) and 1−3 months averaging above 10 °C (50 °F). At least three times as much precipitation in the wettest month of winter as in the driest month of summer, and driest month of summer receives less than 30 mm (1.2 in).

- *Dsd* = Mediterranean-influenced extremely cold subarctic climate; coldest month averaging below −38 °C (−36.4 °F) and 1−3 months averaging above 10 °C (50 °F). At least three times as much precipitation in the wettest month of winter as in the driest month of summer, and driest month of summer receives less than 30 mm (1.2 in).

Group E: Polar and Alpine (Montane) Climates

This type of climate has every month of the year with an average temperature below 10 °C (50 °F).

- *ET* = Tundra climate; average temperature of warmest month between 0 °C (32 °F) and 10 °C (50 °F).

- *EF* = Ice cap climate; eternal winter, with all 12 months of the year with average temperatures below 0 °C (32 °F).

Group A: Tropical/Megathermal Climates

Tropical climates are characterized by constant high temperatures (at sea level and low elevations); all 12 months of the year have average temperatures of 18 °C (64.4 °F) or higher. They are subdivided as follows:

Af: Tropical Rainforest Climate

All 12 months have an average precipitation of at least 60 mm (2.4 in). These climates usually occur within 10° latitude of the equator. This climate has no natural seasons in terms of thermal and moisture changes. When it is dominated most of the year by the doldrums low-pressure system due to the presence of the Intertropical Convergence Zone (ITCZ) and when they are no cyclones then the climate is qualified as equatorial. When the trade winds are dominant most of the year, the climate is a tropical trade-wind rainforest climate.

Some of the places with this climate are indeed uniformly and monotonously wet throughout the year (e.g., the northwest Pacific coast of South and Central America, from Ecuador to Costa Rica; see, for instance, Andagoya, Colombia), but in many cases, the period of higher sun and longer days is distinctly wettest (as at Palembang, Indonesia) or the time of lower sun and shorter days may have more rain (as at Sitiawan, Malaysia). Among these places some have a pure equatorial climate (Balikpapan, Kuala Lumpur, Kuching, Lae, Medan, Paramaribo, Pontianak and Singapore) with the dominant ITCZ aerological mechanism and no cyclones or a subequatorial climate with occasional cyclones (Davao, Ratnapura, Victoria).

(The term aseasonal refers to the lack in the tropical zone of large differences in daylight hours and mean monthly (or daily) temperature throughout the year. Annual cyclic changes occur in the tropics, but not as predictably as those in the temperate zone, albeit unrelated to temperature, but to water availability whether as rain, mist, soil, or ground water. Plant response (e. g., phenology), animal (feeding, migration, reproduction, etc.), and human activities (plant sowing, harvesting, hunting, fishing, etc.) are tuned to this 'seasonality'. Indeed, in tropical South America and Central America, the 'rainy season' (and the 'high water season') is called *invierno* or *inverno*, though it could occur in the Northern Hemisphere summer; likewise, the 'dry season' (and 'low water season') is called *verano* or *verão*, and can occur in the Northern Hemisphere winter).

Am: Tropical Monsoon Climate

This type of climate results from the monsoon winds which change direction according to the seasons. This climate has a driest month (which nearly always occurs at or soon after the "winter" solstice for that side of the equator) with rainfall less than 60 mm (2.4 in), but at least.

Aw/As: Tropical Savanna Climate

Aw: Tropical Savanna Climate with Non-seasonal or Dry-winter Characteristics

Aw climates have a pronounced dry season, with the driest month having precipitation less than 60 mm (2.4 in) and less than.

Most places that have this climate are found at the outer margins of the tropical zone from the low teens to the mid-20s latitudes, but occasionally an inner-tropical location (e.g., San Marcos, Antioquia, Colombia) also qualifies. Actually, the Caribbean coast, eastward from the Gulf of Urabá on the Colombia–Panamá border to the Orinoco River delta, on the Atlantic Ocean (about 4,000 km), have long dry periods (the extreme is the *BSh* climate, characterised by very low, unreliable precipitation, present, for instance, in extensive areas in the Guajira, and Coro, western Venezuela, the northernmost peninsulas in South America, which receive <300 mm total annual precipitation, practically all in two or three months).

This condition extends to the Lesser Antilles and Greater Antilles forming the circum-Caribbean dry belt. The length and severity of the dry season diminishes inland (southward); at the latitude of the Amazon River—which flows eastward, just south of the equatorial line—the climate is *Af*. East from the Andes, between the dry, arid Caribbean and the ever-wet Amazon are the Orinoco River's llanos or savannas, from where this climate takes its name.

As: Tropical Savanna Climate with Dry-summer Characteristics

Sometimes *As* is used in place of *Aw* if the dry season occurs during the time of higher sun and longer days (during summer). This is the case in parts of Hawaii, northwestern Dominican Republic, East Africa, and the Brazilian Northeastern Coast. In most places that have tropical wet and dry climates, however, the dry season occurs during the time of lower sun and shorter days because of rain shadow effects during the 'high-sun' part of the year.

Group B: Dry (Desert and Semi-arid) Climates

These climates are characterized by actual precipitation less than a threshold value set equal to the potential evapotranspiration. The threshold value (in millimeters) is determined as:

Multiply the average annual temperature in °C by 20, then add (a) 280 if 70% or more of the total precipitation is in the high-sun half of the year (April through September in the Northern Hemisphere, or October through March in the Southern), or (b) 140 if 30%–70% of the total precipitation is received during the applicable period, or (c) 0 if less than 30% of the total precipitation is so received.

According to the modified Köppen classification system used by modern climatologists, total precipitation in the warmest six months of the year is taken as reference instead of the total precipitation in the high-sun half of the year.

If the annual precipitation is less than 50% of this threshold, the classification is *BW* (arid: desert climate); if it is in the range of 50%–100% of the threshold, the classification is *BS* (semi-arid: steppe climate).

A third letter can be included to indicate temperature. Originally, *h* signified low-latitude climate (average annual temperature above 18 °C) while *k* signified middle-latitude climate (average annual temperature below 18 °C), but the more common practice today, especially in the United States, is to use *h* to mean the coldest month has an average temperature above 0 °C (32 °F) (or −3 °C (27 °F)), with *k* denoting that at least one month averages below 0 °C.

Desert areas situated along the west coasts of continents at tropical or near-tropical locations characterized by frequent fog and low clouds, despite the fact that these places rank among the driest on earth in terms of actual precipitation received are labelled *BWn* with the n denoting a climate characterized by frequent fog. The *BSn* category can be found in foggy coastal steppes.

BW: Arid Climate

• Coober Pedy, Australia (*BWh*)	• Dubai, United Arab Emirates (*BWh*)
• Alice Springs, Australia (*BWh*)	• Muscat, Oman (*BWh*)
• Almería, Andalusia, Spain (*BWh*, bordering on *BSh*)	• Cairo, Egypt (*BWh*)
• Las Palmas, Canary Islands, Spain (*BWh*)	• Alexandria, Egypt (*BWh*)
• Baghdad, Iraq (*BWh*)	• Khartoum, Sudan (*BWh*)
• Upington, Northern Cape South Africa (*BWh*)	• Djibouti City, Djibouti (*BWh*)
• Hermosillo, Sonora, Mexico (*BWh*)	• Nouakchott, Mauritania (*BWh*)
• Phoenix, Arizona, United States (*BWh*)	• Timbuktu, Mali (*BWh*)
• Las Vegas, Nevada, United States (*BWh*)	• Lima, Peru (*BWh*)
• Death Valley, California, United States (*BWh*), location of the hottest air temperature ever recorded on Earth.	• Ashgabat, Turkmenistan (*BWk*)
	• Turpan, Xinjiang, China (*BWk*)
• Eilat, Southern District, Israel (*BWh*)	• Leh, India (*BWk*)
• ʻAziziya, Jafara, Libya (*BWh*)	• Nukus, Uzbekistan (*BWk*)
• Karachi, Pakistan (*BWh*)	• El Paso, Texas, United States (*BWk*, bordering on *BWh*)
• Doha, Qatar (*BWh*)	• Aral, Kazakhstan (*BWk*)
• Mecca, Makkah Region, Saudi Arabia (*BWh*)	• Damascus, Syria (*BWk*)
• Riyadh, Saudi Arabia (*BWh*)	• Walvis Bay, Erongo Region, Namibia (*BWk*)
• Kuwait City, Capital Governorate, Kuwait (*BWh*)	

BS: Semi-arid (Steppe) Climate

• Lahore, Punjab, Pakistan (*BSh*)	• Yerevan, Armenia (*BSk*, bordering on *Dfa*)
• Jodhpur, India (*BSh*)	• Baku, Azerbaijan (*BSk*)
• Oranjestad, Aruba (*BSh*)	• Denver, Colorado, United States (*BSk*)
• Alicante, Spain (*BSh*)	• Casper, Wyoming, United States (*BSk*)
• Santa Cruz de Tenerife, Canary Islands, Spain (*BSh*)	• Butte, Montana, United States (*BSk*)
• Mount Isa, Queensland, Australia (*BSh*)	• Zaragoza, Spain (*BSk*)
• Honolulu, Hawaii, United States (*BSh*)	• Konya, Turkey (*BSk*)
• Odessa, Texas, United States (*BSh*)	• Zacatecas City, Zacatecas, Mexico (*BSk*)
• Monterrey, Mexico, (*BSh*)	• Astrakhan, Russia (*BSk*)
• Querétaro City, Querétaro, Mexico (*BSh*)	• Ulan-Ude, Russia (*BSk*)
• Maracaibo, Venezuela (*BSh*)	• Lethbridge, Alberta, Canada (*BSk*)
• Petrolina, Pernambuco, Brazil (*BSh*)	• Moose Jaw, Saskatchewan, Canada (*BSk*, bordering on *Dfb*)
• Patos, Paraíba, Brazil (*BSh*)	• Saskatoon, Saskatchewan, Canada (*BSk*, bordering on *Dfb*)
• Piraeus, Greece (*BSh*)	• Kabul, Afghanistan (*BSk*)
• Nicosia, Cyprus (*BSh*)	• Aleppo, Syria (*BSk*)
• Amman, Amman Governorate, Jordan (*BSh*)	• Tianjin, China (*BSk* bordering on *Dwa*)
• Tripoli, Libya (*BSh*)	• Shijiazhuang, Hebei, China (*BSk*)
• Mogadishu, Somalia (*BSh*)	• Lhasa, Tibet Autonomous Region, China (*BSk* bordering on *Dwb*)
• Dakar, Senegal (*BSh*)	• Ulaanbaatar, Mongolia (*BSk* bordering on *Dwb*)
• Ouagadougou, Burkina Faso (*BSh*)	• Comodoro Rivadavia, Argentina (*BSk*)
• Niamey, Niger (*BSh*)	• La Quiaca, Jujuy, Argentina (*BSk*)
• Airolaf, Djibouti (*BSh*)	• L'Agulhas, Western Cape, South Africa (*BSk*)
• N'Djamena, Chad (*BSh*)	• Sana'a, Yemen (*BSk*)
• Luanda, Angola (*BSh*)	• Asmara, Eritrea (*BSk*)
• Windhoek, Namibia (*BSh*)	

Group C: Temperate/Mesothermal Climates

In the Koppen climate system, temperate climates are defined as having an average temperature above 0 °C (32 °F) (or −3 °C (26.6 °F) in their coldest month but below 18 °C (64.4 °F). The average temperature of −3 °C (26.6 °F) roughly coincides with the equatorward limit of frozen ground and snowcover lasting for a month or more.

The second letter indicates the precipitation pattern—*w* indicates dry winters (driest winter month average precipitation less than one-tenth wettest summer month average precipitation. *s* indicates at least three times as much rain in the wettest month of winter as in the driest month of summer. *f* means significant precipitation in all seasons (neither above-mentioned set of conditions fulfilled).

The third letter indicates the degree of summer heat—*a* indicates warmest month average temperature above 22 °C (71.6 °F) while *b* indicates warmest month averaging below 22 °C but with at least four months averaging above 10 °C (50.0 °F), and *c* indicates one to three months averaging above 10 °C (50.0 °F).

Csa: Mediterranean Hot Summer Climates

These climates usually occur on the western sides of continents between the latitudes of 30° and 45°. These climates are in the polar front region in winter, and thus have moderate temperatures and changeable, rainy weather. Summers are hot and dry, due to the domination of the subtropical high pressure systems, except in the immediate coastal areas, where summers are milder due to the nearby presence of cold ocean currents that may bring fog but prevent rain.

Csb: Mediterranean Warm/Cool Summer Climates

Dry-summer climates sometimes extend to additional areas (sometimes well north or south of) typical Mediterranean climates, however since their warmest month average temperatures do not reach 22 °C (71.6 °F) they are classified as *Csb*. Some of these areas would border the oceanic climate (*Cfb*), except their dry-summer patterns meet Köppen's *Cs* minimum thresholds.

Csc: Mediterranean Cold Summer Climates

Cold summer Mediterranean climates (*Csc*) exist in high-elevation areas adjacent to coastal *Csb* climate areas, where the strong maritime influence prevents the average winter monthly temperature from dropping below 0 °C. This climate is rare and is predominantly found in climate fringes and isolated areas of the Cascades and Andes Mountains, as the dry-summer climate extends further poleward in the Americas than elsewhere. Rare instances of this climate can be found in some coastal locations in the North Atlantic and at high altitudes in Hawaii.

Cfa: Humid Subtropical Climates

These climates usually occur on the eastern coasts and eastern sides of continents, usually in the high 20s and 30s latitudes. Unlike the dry summer Mediterranean climates, humid subtropical climates have a warm and wet flow from the tropics that creates warm and moist conditions in the summer months. As such, summer (not winter as is the case in Mediterranean climates) is often the wettest season.

The flow out of the subtropical highs and the summer monsoon creates a southerly flow from the tropics that brings warm and moist air to the lower east sides of continents. This flow is often what brings the frequent but short-lived summer thundershowers so typical of the more southerly subtropical climates like the far southern United States, southern China and Japan.

Cfb: Oceanic Climate

Marine West Coast Climate

Cfb climates usually occur in the higher middle latitudes on the western sides of continents between the latitudes of 40° and 60°; they are typically situated immediately poleward of the Mediterranean climates, although in Australia and extreme southern Africa this climate is found immediately poleward of temperate climates, and at a somewhat lower latitude. In western Europe, this climate occurs in coastal areas up to 63°N in Norway.

These climates are dominated all year round by the polar front, leading to changeable, often overcast weather. Summers are mild due to cool ocean currents, although hotter, stable weather patterns can set in for periods of time, typically longer at the lower latitude range in the northern hemisphere. Winters are milder than other climates in similar latitudes, but usually very cloudy however not always wet. *Cfb* climates are also encountered at high elevations in certain subtropical and tropical areas, where the climate would be that of a subtropical/tropical rain forest if not for the altitude. These climates are called "highlands".

Subtropical Highland Climate with Uniform Rainfall

Subtropical highland climates with uniform rainfall (*Cfb*) are a type of oceanic climate mainly found in highlands of Australia, such as in or around the Great Dividing Range in the states of New South Wales and Victoria, and also sparsely in other continents, such as in South America, among others. Unlike a typical *Cwb* climate, they tend to have rainfall spread evenly throughout the year. They have characteristics of both the *Cfb* and *Cfa* climates, but unlike these climates, they have a high diurnal temperature variation and low humidity, owing to their inland location and relatively high elevation.

Cfc: Subpolar Oceanic Climate

Subpolar oceanic climates (*Cfc*) occur poleward of or at higher elevations than the maritime temperate climates, and are mostly confined either to narrow coastal strips on the western poleward margins of the continents, or, especially in the Northern Hemisphere, to islands off such coasts. They occur in both hemispheres, most often at latitudes from 60° north and south to 70° north and south.

Cwa: Dry-winter Humid Subtropical Climate

Cwa is monsoonal influenced, having the classic dry winter – wet summer pattern associated with tropical monsoonal climates.

Cwb: Dry-winter Subtropical Highland Climate

Dry-winter subtropical highland climate (*Cwb*) is a type of climate mainly found in highlands inside the tropics of Central America, South America, Africa and Asia or areas in the subtropics. Winters are noticeable and dry, and summers can be very rainy. In the tropics, the rainy season is provoked by the tropical air masses and the dry winters by subtropical high pressure.

Cwc: Dry-winter Subpolar Oceanic Climate

Found located mainly above *Cwb* climates, it is found mainly in isolated locations mostly in the Andes in Bolivia, Peru and Argentina, as well as in sparse mountain locations in Southeast Asia. It is not common.

- El Alto, Bolivia (*Cwc*)

- Puno, Peru (*Cwc*)

Group D: Continental Microthermal Climates

The snowy city of Sapporo.

These climates have an average temperature above 10 °C (50 °F) in their warmest months, and a coldest month average below 0 °C (or −3 °C (27 °F). These usually occur in the interiors of continents and on their upper east coasts, normally north of 40°N. In the Southern Hemisphere, group D climates are extremely rare due to the smaller land masses in the middle latitudes and the almost complete absence of land at 40–60°S, existing only in some highland locations.

Dfa/Dwa/Dsa: Hot Summer Continental Climates

Dfa climates usually occur in the high 30s and low 40s latitudes, with a qualifying average temperature in the warmest month of greater than 22 °C/72 °F. In Europe, these climates tend to be much drier than in North America. *Dsa* exists at higher elevations adjacent to areas with hot summer Mediterranean (*Csa*) climates.

These climates exist only in the northern hemisphere because the southern hemisphere has no locations that get the combination of hot summers and snowy winters because the southern hemisphere has no large landmasses isolated from the moderating effects of the sea within the upper-middle latitudes.

In eastern Asia, *Dwa* climates extend further south due to the influence of the Siberian high pressure system, which also causes winters there to be dry, and summers can be very wet because of monsoon circulation.

Dsa exists only at higher elevations adjacent to areas with hot summer Mediterranean (*Csa*) climates.

Dfb/Dwb/Dsb: Warm Summer Continental or Hemiboreal Climates

Dfb climates are immediately poleward of hot summer continental climates, generally in the high 40s and low 50s latitudes in North America and Asia, and also extending to higher latitudes in central and eastern Europe and Russia, between the maritime temperate and continental subarctic climates, where it extends up to 65 degrees latitude in places.

Dfb Examples:

• Kushiro, Hokkaido, Japan (*Dfb*)	• Warsaw, Poland (*Dfb*)
• Karaganda, Kazakhstan (*Dfb*)	• Pristina, Kosovo (*Dfb*)
• Oslo, Norway (*Dfb*)	• Erzurum, Turkey (*Dfb*)
• Lillehammer, Norway (*Dfb*)	• Ardahan, Turkey (*Dfb*)
• Stockholm, Sweden (*Dfb*)	• Edmonton, Alberta, Canada (*Dfb*)
• Helsinki, Finland (*Dfb*)	• Winnipeg, Manitoba, Canada (*Dfb*)
• Tallinn, Estonia (*Dfb*)	• Ottawa, Ontario, Canada (*Dfb*)
• Riga, Latvia (*Dfb*)	• Quebec City, Quebec, Canada (*Dfb*)
• Vilnius, Lithuania (*Dfb*)	• Moncton, New Brunswick, Canada (*Dfb*)
• Kiev, Ukraine (*Dfb*)	• Halifax, Nova Scotia, Canada (*Dfb*)
• Moscow, Russia (*Dfb*)	• Buffalo, New York, United States (*Dfb*, bordering on *Dfa*)
• Novosibirsk, Russia (*Dfb*)	• Marquette, Michigan, United States (*Dfb*)
• Saint Petersburg, Russia (*Dfb*)	• Portland, Maine, United States (*Dfb*)
• Minsk, Belarus (*Dfb*)	

Dwb Examples:

- Calgary, Alberta, Canada (*Dwb*, bordering on *Dfb*),

- Heihe, China (*Dwb*),

- Vladivostok, Russia (*Dwb*),

- Irkutsk, Russia (*Dwb*, bordering on *Dwc*),

- Baruunturuun, Mongolia (*Dwb*),

- Pyeongchang County, South Korea (*Dwb*),

- Pembina, North Dakota, United States (*Dwb*).

Dsb arises from the same scenario as *Dsa*, but at even higher altitudes or latitudes, and chiefly in North America, since the Mediterranean climates extend further poleward than in Eurasia.

Dfc/Dwc/Dsc: Subarctic or Boreal Climates

Dfc, *Dsc* and *Dwc* climates occur poleward of the other group D climates, or at higher altitudes, generally between the 55° to 65° North latitudes, occasionally reaching up to the 70° N latitude.

Dfd/Dwd/Dsd: Subarctic or Boreal Climates with Severe Winters

Places with this climate have severe winters, with the temperature in their coldest month lower than −38 °C. These climates occur only in eastern Siberia. The names of some of the places with this climate have become veritable synonyms for extreme, severe winter cold.

Examples:

- Yakutsk, Sakha Republic, Russia (*Dfd*),
- Oymyakon, Sakha Republic, Russia (*Dfd, bordering on Dwd*),
- Verkhoyansk, Sakha Republic, Russia (*Dfd*),
- Allakh-Yun, Sakha Republic, Russia (*Dwd*),
- Ust-Nera, Sakha Republic, Russia (*Dwd*),
- Zhigansk, Sakha Republic, Russia (*Dsd*, bordering on *Dfd*).

Group E: Polar Climates

In the Köppen climate system, polar climates are defined as the warmest temperature of any month is below 10 °C (50 °F). Polar climates are further divided into two types, tundra climates and icecap climates:

ET: Tundra Climate

Tundra climate (*ET*): Warmest month has an average temperature between 0 and 10 °C. These climates occur on the northern edges of the North American and Eurasian land masses (generally north of 70 °N although it may be found farther south depending on local conditions), and on nearby islands. *ET* climates are also found on some islands near the Antarctic Convergence, and at high elevations outside the polar regions, above the tree line.

These *ET* climates are a colder and more continental variants of tundra. They would have characteristics of the ice cap climate, but still manage to see monthly average temperatures above 0 °C (32 °F):

EF: Ice Cap Climate

Ice cap climate (*EF*): This climate is dominant in Antarctica and inner Greenland, but also occurs at extremely high altitudes on mountains, above even tundra. Monthly average temperatures never exceed 0 °C (32 °F).

Ecological Significance

The Köppen climate classification is based on the empirical relationship between climate and vegetation. This classification provides an efficient way to describe climatic conditions defined by temperature and precipitation and their seasonality with a single metric. Because climatic conditions identified by the Köppen classification are ecologically relevant, it has been widely used to map geographic distribution of long term climate and associated ecosystem conditions.

Over the recent years, there has been an increasing interest in using the classification to identify changes in climate and potential changes in vegetation over time. The most important ecological significance of the Köppen climate classification is that it helps to predict the dominant vegetation type based on the climatic data and vice versa.

In 2015, a Nanjing University paper published in *Nature* analyzing climate classifications found that between 1950 and 2010, approximately 5.7% of all land area worldwide had moved from wetter and colder classifications to drier and hotter classifications. The authors also found that the change "cannot be explained as natural variations but are driven by anthropogenic factors."

Trewartha Climate Classification Scheme

The Trewartha climate classification is a climate classification system published by American geographer Glenn Thomas Trewartha in 1966, and updated in 1980. It is a modified version of the 1899 Köppen system, created to answer some of the deficiencies of the Köppen system. The Trewartha system attempts to redefine the middle latitudes to be closer to vegetation zoning and genetic climate systems. It was considered a more true or "real world" reflection of the global climate.

For example, under the standard Köppen system, in the United States, western Washington and Oregon are classed into the same climate zone as southern California, even though the two regions have strikingly different weather and vegetation. Under the old Köppen system cool oceanic climates like that of London or Seattle were classed in the same zone as hot subtropical cities like Savannah, Georgia or Brisbane, Australia. In the United States, locations like Colorado and Kansas, which have long, severe winter climates where plants are completely dormant, were classed into the same climate zone as Louisiana or northern Florida which have mild winters and a green winter landscape.

Tectonic–climatic Interaction

Tectonic–climatic interaction is the interrelationship between tectonic processes and the climate system. The tectonic processes in question include orogenesis, volcanism, and erosion, while relevant climatic processes include atmospheric circulation, orographic lift, monsoon circulation and the rain shadow effect. As the geological record of past climate changes over millions of years is sparse and poorly resolved, many questions remain unresolved regarding the nature of tectonic-climate interaction, although it is an area of active research by geologists and palaeoclimatologists.

Orographic Controls on Climate

Depending on the vertical and horizontal magnitude of a mountain range, it has the potential to have strong effects on global and regional climate patterns and processes including: deflection of atmospheric circulation, creation of orographic lift, altering monsoon circulation, and causing the rain shadow effect.

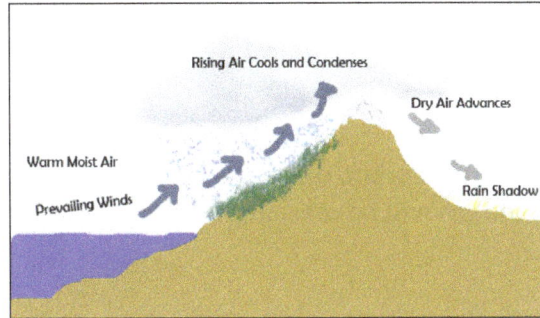

Simple illustration of the rain shadow effect.

One example of an elevated terrain and its effect on climate occurs in the Southeast Asian Himalayas, the world's highest mountain system. A range of this size has the ability to influence geographic temperature, precipitation, and wind. Theories suggest that the uplift of the Tibetan Plateau has resulted in stronger deflections of the atmospheric jet stream, a heavier monsoonal circulation, increased rainfall on the front slopes, greater rates of chemical weathering, and thus lower atmospheric CO_2 concentrations. It is possible that the spatial magnitude of this range is so great that it creates a regional monsoon circulation in addition to disrupting hemispheric-scale atmospheric circulation.

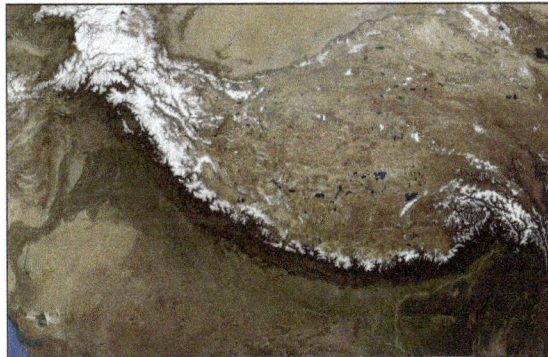

Example of the rain shadow effect in the Himalayas.

The monsoon season in Southeast Asia occurs due to the Asian continent becoming warmer than the surrounding oceans during the summer; as a low-pressure cell is created above the continents, a high-pressure cell forms over the cooler ocean, causing advection of moist air, creating heavy precipitation from Africa to Southeast Asia. However, the intensity of the rainfall over Southeast Asia is greater than the African monsoon, which can be attributed to the awesome size of the Asian continent compared to the African continent and the presence of a vast mountain system. This not only affects the climate of Southeast Asia, but modifies the climate in neighboring areas such as Siberia, central Asia, the Middle East, and the Mediterranean basin as well. To test this a model was created that changed only the topography of current landmasses, which resulted in correlations between the model and global fluctuations in precipitation and temperature over the past 40 Myr. interpreted by scientists.

It is commonly agreed upon that global climate fluctuations are strongly dictated by the presence or absence of greenhouse gases in the atmosphere and carbon dioxide (CO_2) is typically considered the most significant greenhouse gas. Observations infer that large uplifts of mountain ranges globally result in higher chemical erosion rates, thus lowering the volume of CO_2 in the atmosphere as well as causing global cooling. This occurs because in regions of higher elevation there are higher rates of mechanical erosion (i.e. gravity, fluvial processes) and there is constant exposure and availability of materials available for chemical weathering. The following is a simplified equation describing the consumption of CO_2 during chemical weathering of silicates:

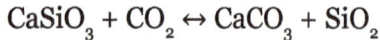

$$CaSiO_3 + CO_2 \leftrightarrow CaCO_3 + SiO_2$$

From this equation, it is inferred that carbon dioxide is consumed during chemical weathering and thus lower concentrations of the gas will be present in the atmosphere as long as chemical weathering rates are high enough.

Climate-driven Tectonism

There are scientists who reject that uplift is the sole cause of climate change and are in favor of uplift as a result of climate change. Some geologists theorize that a cooler and stormier climate (such as glaciations and increased precipitation) can give a landscape a younger appearance such as incision of high terrains and increased erosion rates. Glaciers are a powerful eroding agent with the ability to incise and carve deep valleys and when rapid erosion of the earth's surface occurs, especially in an area of limited relief, it is possible for isostatic rebound to occur, creating high peaks and deep valleys. A lack of glaciation or precipitation can cause an increase in erosion, but can vary between localities. It is possible to create erosion in the absence of precipitation because there would be a decrease in vegetation, which typically acts as a protective cover for the bedrock.

Peaks and valleys of the Torres del Paine range of the Andes in Chile.

Models also suggest that certain topographic features of the Himalayan and Andes region are determined by an erosional/climatic interaction as opposed to tectonism. These models reveal a correlation between regional precipitation and a maximum topographic limit at the plateau margin. In the southern Andes where there is relatively low precipitation and denudation rates, there is no real extreme topography present at the plateau margin while in the north there are higher rates of precipitation and the presence of extreme topography.

Another interesting theory comes from an investigation of the uplift of the Andes during the Cenozoic. Some scientists hypothesize that the tectonic processes of plate subductionand mountain building are products of erosion and sedimentation. When there is an arid climate influenced by the rain shadow effect in a mountainous region, sediment supply to the trench can be reduced or even cut off. These sediments are thought to act as lubricants at the plate interface and this reduction increases the shear stress present at the interface that is large enough to support the high Andes.

Volcanism

Around the world, dotting the map are volcanoes of all shapes and sizes. Lining the landmass around the Pacific Ocean are the well-known volcanoes of the Pacific Ring of Fire. From the Aleutian Islands to the Andes Mountains in Chile, these volcanoes have sculpted their local and regional environments. Aside from admiring their majestic beauty, one might wonder how these geologic wonders work and what role they play in changing the landscape and atmosphere. Principally, volcanoes are geologic features that exude magmatic material from below Earth's surface onto the surface. Upon reaching the surface, the term "magma" disappears and "lava" becomes the common nomenclature. This lava cools and forms igneous rock. By examining igneous rocks, it is possible to derive a chain of events that led from the original melt of the magma to the crystallization of the lava at Earth's surface. By examining igneous rocks, it is possible to postulate evidence for volcanic outgassing, which is known to alter atmospheric chemistry. This alteration of atmospheric chemistry changes climate cycles both globally and locally.

Fundamentals of Igneous Rock and Magmatic Gas Formation

Magmas are the starting point for the creation of a volcano. In order to understand volcanism, it is critical to understand the processes that form volcanoes. Magmas are created by keeping temperature, pressure, and composition (known as P-T-X) in the realm of melt conditions. The pressure and temperature for melts are understood by knowing the chemistry of the melt. To keep magma in a melt condition, a change in one variable will result in the change of another variable in order to maintain equilibrium (i.e. Le Chatlier's Principle). The production of magma is accomplished in multiple ways: 1) subduction of oceanic crust, 2) creation of a hot spot from a mantle plume, and 3) divergence of oceanic or continental plates. The subduction of oceanic crust produces a magmatic melt usually at great depth. Yellowstone National Park is a hot spot located within the center of a continent. Divergence of continental plates (i.e. the Atlantic Mid-Ocean ridge complex) creates magmas very near the surface of the Earth. A plume of heat from the mantle will melt rocks, creating a hot spot, which can be located at any depth in the crust. Hot spots in oceanic crust develop different magmatic plumbing systems based on plate velocities. Hawaii and the Madeira Archipelago (off the West coast of Africa) are examples of volcanic complexes with two different plumbing systems. Because islands such as Hawaii move more quickly than Madeira, the layered rocks at Hawaii have a different chemistry than those at Madeira.The layers beneath Hawaii and Madeira are different because the magma produced underground at these locations rests for different amounts of time. The longer the amount of time magma will rest underground, the warmer the host rocks become. Fractionation of crystals from melt is partially driven by heat; therefore, the igneous rock produced will vary from an insulated host to a non-insulated host. Each of these avenues of magmatic creation develops different igneous rocks and, thus, various P-T-X histories.

In order to understand the creation of igneous rocks from a melt, it is fundamental to understand the concepts produced by Drs. Norman Bowen and Frank Tuttle from the $NaAlSiO_4$-$KAlSiO_4$-SiO_2-H_2O system. Tuttle and Bowen accomplished their work by using experimental petrologic laboratories that produce synthetic igneous materials from mixes of reagents. Observations from these experiments indicate that as a melt cools, it will produce derivative magmas and igneous rock. Following Bowen's research, the magma will crystallize a mafic igneous rock prior to a felsic igneous rock. As this crystallization process occurs in nature, pressure and temperature decrease, which changes the composition of the melt along various stages of the process. This constantly changing chemical environment alters the final composition that reaches the Earth's surface.

The evolution of magmatic gases depends on the P-T-X history of the magma. These factors include the composition of assimilated materials and composition of parent rock. Gases develop in magma through two different processes: first and second boiling. First boiling is defined as a decrease in confining pressure below the vapor pressure of the melt. Second boiling is defined as an increase in vapor pressure due to crystallization of the melt. In both cases, gas bubbles exsolve in the melt and aid the ascent of the magma towards the surface. As the magma ascends towards the surface, the temperature and confining pressure decrease. A decrease in temperature and confining pressure will allow an increase in crystallization and vapor pressure of the dissolved gas. Depending on the composition of the melt, this ascent can be either slow or fast. Felsic magmas are very viscous and travel to the surface of the Earth slower than mafic melts whose silica levels are lower. The amount of gas available to be exsolved and the concentrations of gases in the melt also control ascension of the magma. If the melt contains enough dissolved gas, the rate of exsolution will determine the magmas rate of ascension. Mafic melts contain low levels of dissolved gases whereas felsic melts contain high levels of dissolved gases. The rate of eruption for volcanoes of different compositions is not the controlling factor of gas emission into the atmosphere. The amount of gas delivered by an eruption is controlled by the origin of the magma, the crustal path the magma travels through, and several factors dealing with P-T-x at the Earth's surface. When felsic melts reach the surface of the Earth, they are generally very explosive (i.e. Mount St. Helens). Mafic melts generally flow over the surface of the Earth and form layers (i.e. Columbia River Basalt). Magma development under continental crust develops a different type of volcano than magmas that are generated under oceanic crust. Subduction zones produce volcanic island arcs (such as the Aleutian Islands, Alaska) and non-arc volcanism (such as Chile and California). Typically, arc volcanism is more explosive than non-arc volcanism due to the concentrations and amounts of gasses withheld in the magma underground.

Fluid inclusion analysis from fluids trapped in minerals can show a path of volatile evolution in volcanic rocks. Isotopic analyses and interpretation of degassing scenarios are required in order to derive the origin of magmatic volatiles. When gas bubbles accumulate in a melt that is crystallizing, they create a vesicular texture. Vesicles are created by super cooling a melt while gases are present. Because the rock crystallized very quickly while in the Earth's atmosphere, it is possible to examine some igneous rocks for fluids trapped in vesicles. By examining many different inclusions, it is possible to detect crustal assimilation and depressurization that account for volatile release.

Methods of Characterizing Igneous Rocks

The methods by which petrologists examine igneous rocks and synthetically produced materials are optical petrography, X-ray diffraction (XRD), electron probe microanalysis (EPMA), laser ablation inductively coupled mass spectrometry (LA-ICP-MS), and many others. Methods such as optical petrography aid the researcher in understanding various textures of igneous rocks and, also, the mineralogical composition of the rock. XRD methods define the mineralogical constituents of the rock being tested; therefore, composition is only known based on the mineralogical composition discovered using this method. EPMA reveals textural features of the rock on the micron level. It also reveals a composition of the rock based on elemental abundance. For information about fluids trapped in an igneous rock, LA-ICP-MS could be used. This is accomplished by finding rocks with small pockets of fluid or vapor, acquiring the fluid or vapor, and testing the fluid or vapor for various elements and isotopes.

Volcanic Emissions and Effects

While most volcanoes emit some mixture of the same few gasses, each volcano's emissions contain different ratios of those gasses. Water vapour (H_2O) is the predominant gas molecule produced, closely followed by carbon dioxide (CO_2) and sulfur dioxide (SO_2), all of which can function as greenhouse gasses. A few unique volcanoes release more unusual compounds. For example, mud volcanoes in Romania belch out much more methane gas than H_2O, CO_2, or SO_2 –95–98% methane (CH_4), 1.5–2.3% CO_2, and trace amounts of hydrogen and helium gas. To measure volcanic gases directly, scientists commonly use flasks and funnels to capture samples directly from volcanic vents or fumaroles. The advantage of direct measurement is the ability to evaluate trace levels in the gaseous composition. Volcanic gasses can be indirectly measured using Total Ozone Mapping Spectrometry (TOMS), a satellite-remote sensing tool which evaluates SO_2 clouds in the atmosphere. TOMS' disadvantage is that its high detection limit can only measure large amounts of exuded gases, such as those emitted by an eruption with a Volcanic Explosivity Index (VEI) of 3, on a logarithmic scale of 0 to 7.

Sulfur ejection from volcanoes has a tremendous impact environmental impact, and is important to consider when studying the large-scale effects of volcanism. Volcanoes are the primary source of the sulfur (in the form of SO_2) that ends up in the stratosphere, where it then reacts with OH radicals to form sulfuric acid (H_2SO_4). When the sulfuric acid molecules either spontaneously nucleate or condense on existing aerosols, they can grow large enough to form nuclei for raindrops and precipitate as acid rain. Rain containing elevated concentrations of SO_2 kills vegetation, which then reduces the ability of the area's biomass to absorb CO_2 from the air. It also creates a reducing environment in streams, lakes, and groundwater. Because of its high reactivity with other molecules, increased sulfur concentrations in the atmosphere can lead to ozone depletion and start a positive warming feedback.

Volcanoes with a felsic melt composition produce extremely explosive eruptions that can inject massive quantities of dust and aerosols high into the atmosphere. These particulate emissions are potent climate forcing agents, and can provoke a wide variety of responses including warming, cooling, and rainwater acidification. The climatic response depends on the altitude of the dust cloud as well as the size and composition of the dust. Some volcanic silicates cooled extremely quickly, created a glassy texture; their dark colour and reflective nature absorb some radiation and reflect the rest. Such volcanic material injected into the stratosphere blocks solar radiation, heating that layer of the atmosphere and cooling the area beneath it. Wind patterns can distribute the dust over vast geographic regions; for example, the 1815 eruption of Tambora in Indonesia produced so much dust that a cooling of 1 degree Celsius was noted as far away as New England, and lasted for several months. Europeans and Americans called its effect "the year without a summer".

Volcanic emissions contain trace amounts of heavy metals, which can affect the hydrosphere when they are injected into the lower reaches of the atmosphere. When large quantities of these emissions are concentrated into a small area they can damage ecosystems, negatively affect agriculture, and pollute water sources.

Materials being emitted from volcanoes typically carry heavy metals in the trace level. When large quantities of these emissions are collected into a small area, the contamination effects become paramount.

The short-term (months-to-years) impacts of volcanism on the atmosphere, climate and environment are strongly controlled by location, timing, flux, magnitude and emission height of sulfur gases. Episodic explosive eruptions represent the principal perturbation to stratospheric aerosol (though the atmospheric effects of sulfur degassing associated with continental flood basalts might well be more profound). In the troposphere, the picture is less clear but a significant part of the global tropospheric sulfate burden may be volcanogenic. Sulfate aerosol influences the Earth's radiation budget by scattering and absorption of shortwave and long-wave radiation, and by acting as cloud condensation nuclei. When they are brought to the boundary layer and Earth's surface, clouds containing volcanic sulfur in both gaseous and aerosol phases can result in profound environmental and health impacts.

Examples of the environmental and health impacts are agricultural loss due to acid rain and particulate shading, damage to ecosystems, and pollution in the hydrosphere. Intensity of a volcanic eruption is a variable controlling the altitude and effect of ejected material. Though larger eruptions occur less often than smaller eruptions, larger eruptions still deliver more particulate matter into the atmosphere. This year round behavior of emitted material yields mild effects on the atmosphere in comparison to larger eruptions. Over time, changes in the composition of smaller scale eruptions yields changes to atmospheric cycles and the global climate. Larger scale eruptions cause changes to the atmosphere immediately, which in turn leads to climatic changes in the immediate vicinity. The larger the volcanic expulsion, the higher the altitude achieved by the ejected silicate materials. Higher altitude injections are caused by larger intensity eruptions. Larger eruptions do not emit as much, on average, as smaller eruptions. This is related to the return period of the eruptions and the amount of ejected material per eruption. "The injection height of sulfur into the atmosphere represents another important determinant of climate impact. More intense eruptions, i.e., those with higher magma discharge rates, are more likely to loft the reactive sulfur gases into the stratosphere where they can generate climatically effective aerosol."

Eruption intensity of a volcano is not the only factor controlling the altitude of particles. The climate surrounding the volcano constrains the impact of the eruption. Models of eruptions that treat climatic variables as controls and hold eruption intensity constant predict particulate emissions, such as volcanic ash and other pyroclastic debris ejected into the atmosphere, in the tropics to reach higher altitudes than eruptions in arid or polar areas. Some of these climatic variables include humidity, aridity, winds, and atmospheric stability. The observation made by the model matches what is seen in nature: volcanoes in tropical climates have greater eruption heights than those in the poles. If there were a widening of the tropics, the number of volcanoes able to produce higher altitude emissions into the atmosphere would increase. Effects on the climate from the increase in airborne silicate material would be substantial because the height of these tropical eruptions will become more prominent with a widening of the tropics leading to more risks such as cooling, pollution, and aircraft disturbances.

The location of a volcano strongly influences the geographic distribution of atmospheric heating and the development of planetary waves that affect air circulation (especially in the northern hemisphere). Another relevant factor is that the height of the tropopause varies with latitude—at the tropics it is around 16–17 km above sea level but descends to 10–11 km at high latitudes. In general terms, an explosive eruption requires a greater intensity (magma discharge rate) to cross the tropopause in the tropics than at mid to polar latitudes. However, there are two factors that limit this effect. The first is that a high-latitude eruption will have a more limited effect than a low-latitude one because further from the tropics there is less solar energy to intercept. Secondly, atmospheric circulation works in a way to limit the effects of high latitude eruptions. A tropical eruption that pumps aerosol into the stratosphere results in localized heating. This increases the temperature difference in the middle atmosphere between the equator and high latitudes, and thereby enhances meridional air flows that spread aerosol into both hemispheres, promoting climate forcing at a worldwide scale. In contrast, volcanic aerosol injected into the stratosphere from high latitude volcanoes will tend to have the opposite effect on the temperature gradient, acting to stagnate meridional air flow. Very little, if any, of the stratospheric aerosol formed as a result of eruption of a high latitude volcano will reach the opposing hemisphere.

Interaction between Glaciation and Volcanoes

Volcanoes do not only affect the climate, they are affected by the climate. During times of glaciation, volcanic processes slow down. Glacial growth is promoted when summer heat is weak and winter cold is enhanced and when glaciers grow larger, they get heavier. This excess weight causes a reverse effect on the magma chamber's ability to produce a volcano. Thermodynamically, magma will dissolve gases more readily when the confining pressure on the magma is greater than the vapor pressure of the dissolved components. Glacial buildup typically occurs at high elevations, which are also the home to most continental volcanoes. Buildup of ice can cause a magma chamber to fail and crystallize underground. The cause of magma chamber failure occurs when the pressure of ice pressing down on Earth is greater than the pressure being exerted on the magma chamber from heat convection in the mantle. Ice core data from glaciers provides insight into past climate. "Oxygen isotopes and the calcium ion record are essential indicators of climatic variability, while peaks in sulfate ions (SO_4) and in electrical conductivity of the ice indicate volcanic aerosol fallout." As seen in ice cores, volcanic eruptions in the tropics and southern hemisphere are not recorded in the Greenland Ice sheets. Fallout from tropical eruptions can be seen at both poles though this takes nearly two years and consists of only sulfuric precipitation. "One of the striking revelations of the ice core record is the evidence for numerous great eruptions, which have not otherwise been recognized in tephra records. One caveat to the approach is that although the dating of the ice core by counting of seasonal layers is fairly robust, it is not fail-safe. The greater the depth from which the core is retrieved, the more likely it is to have suffered deformation Prevailing winds and atmospheric chemistry play a large role in moving volcanic volatiles from their source to their final locations at the surface or in the atmosphere."

Cretaceous Climate

During the Cretaceous, Earth experienced an unusual warming trend. Two explanations for this warming are attributed to tectonic and magmatic forces. One of the theories is a magmatic super plume inducing a high level of CO_2 into the atmosphere. Carbon dioxide levels in the Cretaceous could have been as high as 3.7 to 14.7 times their present amounts today causing an average 2.8 to 7.7 degrees Celsius.Tectonically, movements of the plates and a sea level fall could cause an additional 4.8 degrees Celsius globally. The combined effect between magmatic and tectonic processes could have placed the Cretaceous Earth 7.6 to 12.5 degrees Celsius higher than today.

A second theory on the warm Cretaceous is the subduction of carbonate materials. By subducting carboniferous materials, a release of carbon dioxide would emit from volcanoes. During the Cretaceous, the Tethys Sea was rich in limestone deposits. By subducting this carboniferous platform, the resulting magma would have become more carbon dioxide rich. Because carbon dioxide dissolves into melts well, it would have remained dissolved until the confining pressure of the magma was low enough to de-gas and release massive quantities of carbon dioxide into the atmosphere causing warming.

Controls of Climate

Climate is the average of weather in that location over a long period of time, usually for at least 30 years. A location's climate can be described by its air temperature, humidity, wind speed and direction, and the type, quantity, and frequency of precipitation. Climate can change, but only over long periods of time. The climate of a region depends on its position relative to many things.

Latitude

The main factor influencing the climate of a region is latitude because different latitudes receive different amounts of solar radiation.

- The equator receives the most solar radiation. Days are equally long year-round and the sun is just about directly overhead at midday.

- The polar regions receive the least solar radiation. The night lasts six months during the winter. Even in summer, the sun never rises very high in the sky. Sunlight filters through a thick wedge of atmosphere, making the sunlight much less intense. The high albedo, because of ice and snow, reflects a good portion of the sun's light.

Atmospheric Circulation

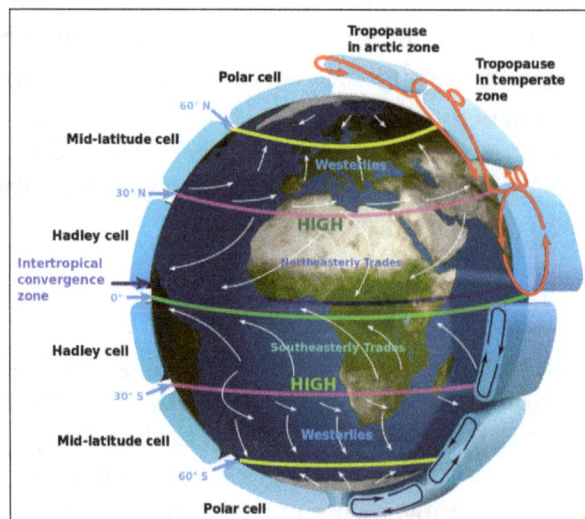

The position of a region relative to the circulation cells and wind belts has a great effect on its climate. In an area where the air is mostly rising or sinking, there is not much wind.

Intertropical Convergence Zone

The Intertropical Convergence Zone (ITCZ) is the low pressure area near the equator in the boundary between the two Hadley Cells. The air rises so that it cools and condenses to create clouds and rain. Climate along the ITCZ is therefore warm and wet. Early mariners called this region the doldrums because their ships were often unable to sail because there were no steady winds.

The ITCZ migrates slightly with the season. Land areas heat more quickly than the oceans. Because there are more land areas in the Northern Hemisphere, the ITCZ is influenced by the heating effect of the land. In Northern Hemisphere summer, it is approximately 5o north of the equator while in the winter it shifts back and is approximately at the equator. As the ITCZ shifts, the major wind belts also shift slightly north in summer and south in winter, which causes the wet and dry seasons in this area.

Hadley Cell and Ferrel Cell Boundaries

At about 30 degrees N and 30 degrees S, the air is fairly warm and dry because much of it came from the equator where it lost most of its moisture at the ITCZ. At this location the air is descending, and sinking air warms and causes evaporation.

Mariners named this region the horse latitudes. Sailing ships were sometimes delayed for so long by the lack of wind that they would run out of water and food for their livestock. Sailors tossed horses and other animals over the side after they died. Sailors sometimes didn't make it either.

Prevailing Winds

The prevailing winds are the bases of the Hadley, Ferrell, and Polar Cells. These winds greatly influence the climate of a region because they bring the weather from the locations they come from. For example, in California, the predominant winds are the westerlies blowing in from the Pacific Ocean, which bring in relatively cool air in summer and relatively warm air in winter. Local winds also influence local climate. For example, land breezes and sea breezes moderate coastal temperatures.

Continental Position

When a particular location is near an ocean or large lake, the body of water plays an extremely important role in affecting the region's climate.

- A maritime climate is strongly influenced by the nearby sea. Temperatures vary a relatively small amount seasonally and daily. For a location to have a true maritime climate, the winds must most frequently come off the sea.

- A continental climate is more extreme, with greater temperature differences between day and night and between summer and winter.

The ocean's influence in moderating climate can be seen in the following temperature comparisons. Each of these cities is located at 37 °N latitude, within the westerly winds.

Maritime climate.

Continental climate.

Ocean Currents

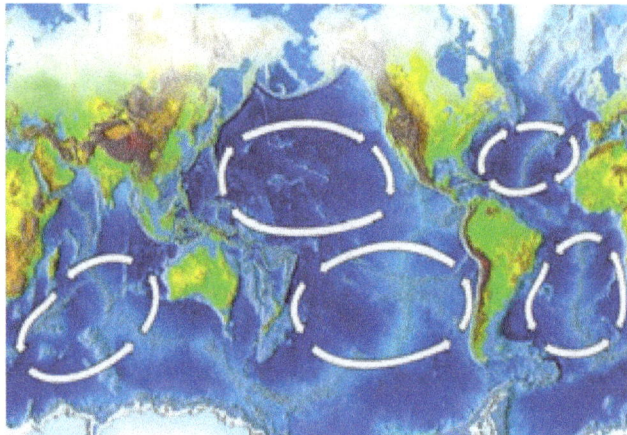

The temperature of the water offshore influences the temperature of a coastal location, particularly if the winds come off the sea. The cool waters of along the western United States is caused by a clockwise rotating ocean current that is bringing cold water from the arctic toward the equator. The climatic effect is that coastal regions of California, Oregon, and Washington are are cool. Coastal upwelling also brings cold, deep water up to the ocean surface off of California, which contributes to the cool coastal temperatures. But that same ocean current brings warm, tropical water to eastern Japan.In the Atlantic Ocean, the northern ocean current, called the Gulf Stream, brings warm water from the tropics to the southern states. This is a major reason why the southern states experience humid conditions in the summer and tornadoes because of all this warm moisture. The Gulf Stream also impacts Europe by bringing warm water northward, making this region that is rather northward warmer than expected.

Altitude and Mountain Ranges

Air pressure and air temperature decreases with altitude. The closer molecules are packed together, the more likely they are to collide. Collisions between molecules give off heat, which warms the air. At higher altitudes, the air is less dense and air molecules are more spread out and less likely to collide. A location in the mountains has lower average temperatures than one at the base of the mountains. In Colorado, for example, Lakewood (5,640 feet) average annual temperature is 62 °F (17 °C), while Climax Lake (11,300 feet) is 42 °F (5.4 °C).

Mountain ranges have two effects on the climate of the surrounding region. The first is something called the rainshadow effect, which brings warm dry climate to the leeward size of a mountain range. The second effect mountains have on climate systems is the ability to separate coastal regions from the rest of the continent. Since a maritime air mass may have trouble rising over a mountain range, the coastal area will have a maritime climate but the inland area on the leeward side will have a continental climate.

References

- Climate, encyclopedia: nationalgeographic.org, Retrieved 13 March, 2019

- tupper, a.; et al. (2009). "tall clouds from small eruptions: the sensitivity of eruption height and fine ash content to tropospheric instability". Natural hazards. 51 (2): 375–401. Doi:10.1007/s11069-009-9433-9

- Urban-climate, science: britannica.com, Retrieved 23 February, 2019

- mcknight, tom l; hess, darrel (2000). "climate zones and types". Physical geography: a landscape appreciation. Upper saddle river, nj: prentice hall. Isbn 978-0-13-020263-5

- Factors-affecting-climate, climate, tutorials-weather-climate, education, what-we-do: ecn.ac.uk, Retrieved 10 July, 2019

- lamb, s; davis, p (2003). "cenozoic climate change as a possible cause for the rise of the andes". Nature. 425 (6960): 792–797. Bibcode:2003natur.425..792l. Doi:10.1038/nature02049. Pmid 14574402

- Controls-of-climate, chapter, geophysical: lumenlearning.com, Retrieved 22 May, 2019

Climatology: Introduction

2

- **Climatic Geomorphology**
- **Climatic Elements**
- **Climatic Zones**

The scientific study of the weather conditions averaged over a period of time is referred to as climatology. It is a branch of atmospheric sciences and a sub-field of physical geography. Some of the elements of weather which are studied within this field are temperature, precipitation and wind. This chapter has been carefully written to provide an easy understanding of the varied facets of climatology.

Climatology is the study of atmospheric conditions over a longer period of time. It includes the study of different kinds of weather that occur at a place. Dynamic change in the atmosphere brings about variation and occasionally great extremes that must be treated on the long term as well as the short term basis. As a result, climatology may be defined as the aggregate of weather at a place over a given time period.

There is diversity of approaches available in climate studies. Figure below illustrates the major subgroups of climatology, the approaches that can be used in their implementation, and the scales at which the work can be completed.

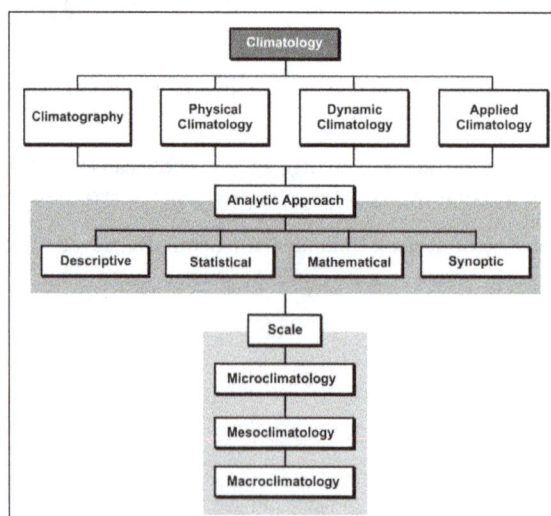

Subgroups, Analytical methods and scales of climatic study.

Climatography consists of the basic presentation of data and its verbal or cartographic description. Physical Climatology deals largely with the energy exchanges and physical components.

Dynamic Climatology is more concerned with atmospheric motion and exchanges that lead to and result from that motion.

Applied Climatology is the scientific application of climatic data to specific problems within such areas of forestry, agriculture, and industry. It can involve the application of climatic data and theory of other disciplines, such as geomorphology and soil science.

The analytical approaches suggested in the figure are self-explanatory, with the possible exception of the synoptic approach, an analytic method that combines each of the others. The object of synoptic climatology is to relate local or regional climates to atmospheric circulation.

Temperature Structure of the Atmosphere

In general terms, the atmosphere can be considered as a series of concentric layers or shells surrounding the earth. The most commonly used method to describe atmospheric layering uses temperature as the variable.

The troposphere is the lowest level, where "weather" occurs. In this layer, there is generally uniform decrease of temperature with height. The lowest part of the troposphere, up to 1.5 km or 2 km is called friction layer. The upper limit of the troposphere is the tropopause. It is zone where generally decrease of temperature ceases and temperature remains fairly constant with height (isothermal-equal temperature). The tropopause also represents the upper limit of large scale turbulence and mixing of the layer.

The stratosphere extends from the 10 km to 45 km above the surface. In this lower section, temperatures are fairly constant, but at an elevation of 30 km they increase toward the upper limit of this zone, the stratospause. Air circulation in the stratosphere is characteristically persistent, with winds blowing at high velocities.

The mesosphere, above the stratopause, is identified by a marked temperature decrease with altitude. Beginning at an elevation of about 80 km, the decline continues until the mesopause is reached.

The thermosphere lying above the mesopause has no defined upper limit. It is so named because a very high thermodynamic temperature attained.

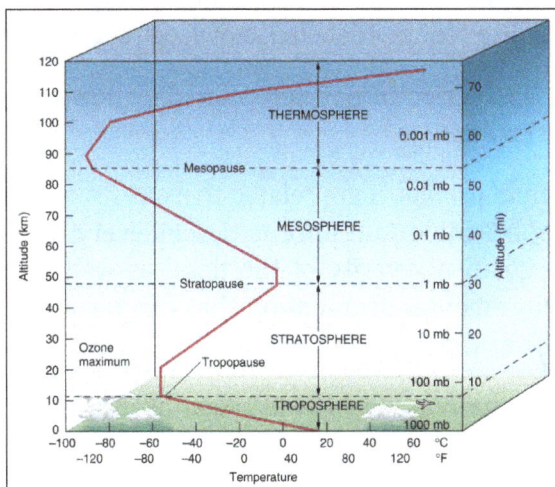

Thermal Structure of Atmosphere.

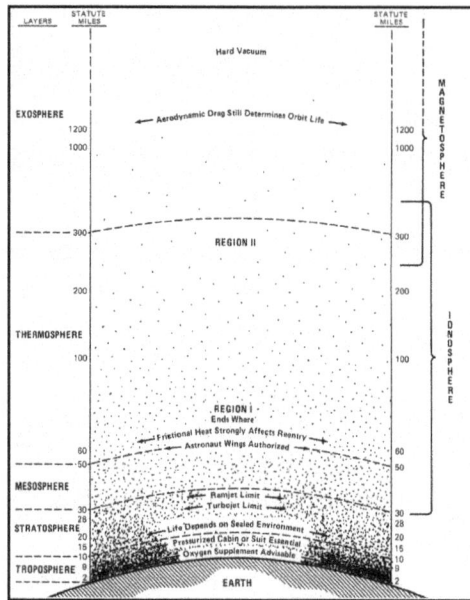

Schematic cross section of Earth's Atmosphere showing broad chemical dividions.

Applications of Climatology

Climatology is a fascinating area of study. It relates directly in which the environment functions and the everyday lives of people in addition to workings and nature of the atmosphere. Applied Climatology is used to:

- Improve efficiency of various economic activities that are influenced by climate,

- Aid in the needs of societal activities,

- Reduce the losses incurred from climatic hazards. Consider the following examples:

  ○ Energy: Climate plays a part in the use and development of both fossil fuels and the renewable energy sources. The use of fossil fuels to generate energy is so climate related that the concept of heating and cooling degree days is used by engineers to estimate energy demands. The potential an area has for the development of renewable sources is directly related to the atmospheric conditions. To evaluate the potential for the direct use of solar energy and wind power requires detailed knowledge of the climatic conditions that prevail.

  ○ Food: The success of agriculture is determined by how well farmers adapt their crops and activities to climate.

  ○ Water: Many human problems are related to excesses or deficits of water. The applied climatologist supplies important information on everything from the chances of flood occurring to the relative severity of the flood as well as his concern with drought includes the environmental degradation that can result when a location experiences slowing increasing aridity.

  ○ Health: Studies of the relationships between health, disease, mortality, and climate show some remarkable results. Climate influences the relative merits of a location as a health resort or as a tourist center. The effect of climates at high altitudes provides an apt example of the hazard.

- ○ Industry and Trade: Applied climatologists have a significant input into the industrial and trade sectors of an economy. For example:

    - Building and construction: site conditions, design in relation to climate, energy costs, construction conditions.

    - Commerce: Storage of materials, accidents, plant operation, sales planning.

    - Communication: Construction, maintenance of systems, appropriate design.

    - Services: Insurance, environmental law, disaster control.

    - Forestry: Productivity, hazards, regeneration, biological hazards.

Climatology is important since it helps determine future climate expectations. Through the use of latitude, one can determine the likelihood of snow and hail reaching the surface. You can also be able to identify the thermal energy from the sun that is accessible to a region. Climatology is the scientific study of climates, which is defined as the mean weather conditions over a period of time. A branch of study within atmospheric sciences, it also takes into account the variables and averages of short-term and long-term weather conditions.

Climatology is different than meteorology and can be divided into different areas of study. Various approaches to this field can be taken, including paleoclimatology, which focuses on studying the climate over the course of the Earth's existence by examining records of tree rings, rocks and sediment, and ice cores. Historical climatology focuses primarily on climate changes throughout history and the effects of the climate on people and events over time. Though both climatology and meteorology are areas of study that are considered branches of similar areas of study, climatology differs from meteorology because its focus is on averages of weather and climatic conditions over a long period of time. Meteorology focuses more on current weather conditions such as humidity, air pressure, and temperatures and forecasting. The mission of the Climatology uses provides a forum for publishing new findings on Environmental principles and technology. Currently our primary research objective is to encourage and assist the development of better and faster measures of Environmental activity. In cases where we believe we can contribute directly, as opposed to through highlighting the work of others, we are producing our own measures of Climatology.

Climatology has developed considerably since the days when it was essentially a climate 'book-keeping' activity, and its status as a credible scientific pursuit was called into question. Despite its centrality to such things as the IPCC process, for some, the nature, scope and methodology of climatology still remain poorly defined. The purpose of this editorial essay, therefore, is to briefly outline for the wider scientific community, the scientific nature and scope of climatology. A secondary aim is to define how and by whom climatology is practised by deriving some simple diagnostics concerning the nature and scope of the subject from papers published in the International Journal of Climatology over the period.

The subject matter of climatology is climate, which is not simply the statistical assemblage of the weather at a location or for a region. Rather it 'is the thermodynamic/hydrodynamic status of the global boundary conditions that determine the concurrent array of weather patterns. This reverses the abstract notion that climate is 'average weather' and suggests that weather is constrained by climate within a 'season's allowable array'. For example, tropical cyclone formation and maintenance is very much constrained by sea surface temperature conditions. Such a definition of climate matches the common usage of the word; for instance, economic, social or political climate is often heard in daily conversations. Used in this way, climate describes the conditions under which 'things' are possible.

Climatology is not only concerned with the analysis of climate patterns and statistics (e.g. temperature, precipitation, atmospheric moisture, atmospheric circulation and disturbances) but also with seasonal to inter-annual climate variability, decadal to millennial climate fluctuations, long-term changes in mean and variability characteristics, climate extremes and seasonality. Climatology also addresses its subject matter on many spatial scales, from the micro through the meso and synoptic to the hemispheric and global. Further, climatology works within a general systems' paradigm. At the heart of this is climate system theory. This states that climate is the manifestation of the interaction among the major climate system components of the atmosphere of hydrosphere, cryosphere, biosphere and land surface and external forcings such as solar variability and long term earth– sun geometry relationships. It also recognises that humans are an integral component of the climate system through their ability to alter levels of atmospheric trace gases. A major goal of climatology is to understand the flow of energy and matter and the feedbacks and non-linear interactions between the main components of the climate system and their associated climate outcomes.

Science is said to be: 'Concerned either with a connected body of demonstrated truths or with observed facts systematically classified and more or less colligated by being brought under general laws, and which includes trustworthy methods for the discovery of new truth within its own domain'.

Clearly, climatology falls within this definition. Some of the demonstrated truths that underpin climatology as a science include:

- Climate is Non-stationary,

- Climate varies over a number of temporal and spatial scales,

- Major modes of atmospheric circulation exist which may produce climate teleconnections,

- Climate is the long-term manifestation of the interaction between the atmosphere and the earth's surface and of processes arising from other causes that are internal and external to the climate system,

- The climate system responds non-linearly to both internal and external forcing and regulates itself through positive and negative feedback,

- The climate of a location is influenced by the balance between large and local scale factors,

- Climate can be a determinant of, a resource for and a hazard to human activities,

- Human activities have the potential to influence climate.

Some of these climatological truths have become codified in the form of theoretical concepts such as ENSO delayed oscillator, energy balance, climate change and climatic attractor theory. Further, a range of methods are utilised for the observation and monitoring of climate, including conventional land-based and upper- air instruments, ocean buoys and profilers and satellite based remote sensing systems. Both statistical and numerical modelling methods are applied to the analysis of data from such observing systems in an attempt not only to uncover new truths about climate but also to offer physically plausible and meaningful explanations for observed phenomena. Moreover, the outcome of statistical analyses of associations and patterns within the climate system often provide the basis for formulating and designing climate modelling experiments. Such experiments are potentially powerful as they offer an insight, from a theoretical perspective, into the sensitivity of the climate system to changing boundary conditions and variations in internal and external forcing.

Climatology concerns itself with the three overarching themes of any science, which are the quest for universality in explanation, the analysis of stability and change and the fundamental importance of empirical data. In posing and answering research questions, climatology has tended to vary between the philosophical dichotomies of positivism and realism, with the former being the dominant one. Positivism connotes an approach based upon observational and experimental evidence that relies on sampling theory, large sample sizes, statistical methods, hypothesis testing and empirical generalisation. This matches the deductive approach to problem solving. In contrast, a realist approach, which predominates in climatology's sister discipline meteorology, places emphasis either on the case study or a small number of cases, the objective being to develop an explanation of the mechanisms that generate the phenomenon under investigation. Intensive process-based case studies can help inform the climatologist about the processes that give rise to general climatic patterns and the variability of these. For example, much of the early case-study-based work on urban energy balances, which helped explain observed temperature distributions in cities and large scale experiments such as the Tropical Oceans Global Atmosphere (TOGA) and Coupled Ocean Atmosphere Response Experiment (COARE) that were designed to shed light on how the ocean and atmosphere are linked, were in this vein.

Climatological literature has traditionally been dominated by the climate outcomes of temperature and precipitation, although the climate of a location or region is the product of the interaction between the surface and a range of thermodynamic and dynamical variables. This is because these two facets of climate are of fundamental importance for a range of human activities and moreover, compared to other climate variables, data availability is much less of a problem. However, as new climate data sets and re-analysis products have become available over the last few decades, new insights into the climatological characteristics of the climate system have emerged. Accordingly, there has been a move away from classification and regional description. As reflected by papers published in the International Journal of Climatology, contemporary climatology is concerned with climate dynamics, climate modelling, seasonal climate prediction satellite rainfall and cloud climatology, land-ocean-atmosphere interactions, land surface climatology, and surface energy balance (including urban) modelling, mechanisms of climatic variability, climate teleconnections and change, local to global atmospheric circulation systems, statistical analysis of climate events including extremes and the construction of homogenous historical climate data series from a variety of sources. Such research areas will evolve rapidly with further developments in local to global scale climate monitoring, statistical and numerical modelling, climate reconstruction techniques and the methods employed to map climate (GIS and geostatistics).

Spurred on by the prospects of climate change and the need to understand the extent to which society at all levels is climate sensitive, climatologists have also become increasingly interested in the impacts of climatic variability and change on a range of human affairs. Climatologists have a critical role to play in working with colleagues outside the physical, engineering and environmental sciences to uncover, explain and understand climate society relationships and undertake climate impact assessments.

Who studies climatology? On the basis of an analysis of the disciplinary affiliation of the first author of papers, published in the International Journal of Climatology over the 20-year period from to, it would appear that a wide range of disciplines undertake climatological research. Dominant among these are university Departments of Geography and Government Organisations. This reflects the traditional interest shown by the discipline of Geography in the workings of the natural environment, the two-way relationship between society and environment and the desire of Governments to understand the nature of the climate system from the point of view of improved prediction and evidence for the potential impacts of climate change. A considerable amount of climatology also originates from national meteorological services and university departments of atmospheric science, meteorology and physics. University research institutes also play a major role in undertaking climatological research. Interestingly, some universities have stand-alone departments of climatology. Although these are usually located in larger faculties of science, this is, nevertheless, an indication that climatology is considered in some places to warrant recognition as a distinct area of activity.

Increasingly, teams of scientists, often with different types of expertise and from a range of institutions, undertake climatological research. This is clearly reflected by a rise in the number of authors contributing to papers published in climatology. In many ways, this reflects the changing nature of the research problems addressed by climate scientists. These necessarily require teams composed of individuals with expertise in database construction, modelling, statistical analysis and interpretation and ecosystem and societal responses to climatic variability and change. Perhaps, more so than other disciplines, climatology is becoming truly interdisciplinary.

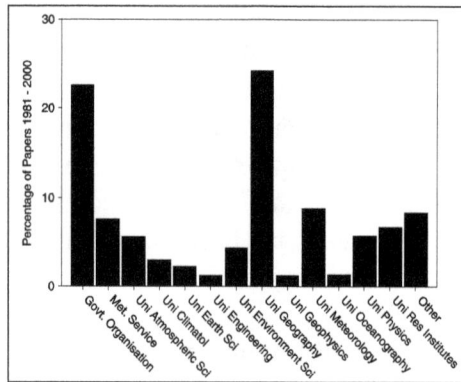

Percentage of papers published by institution type.

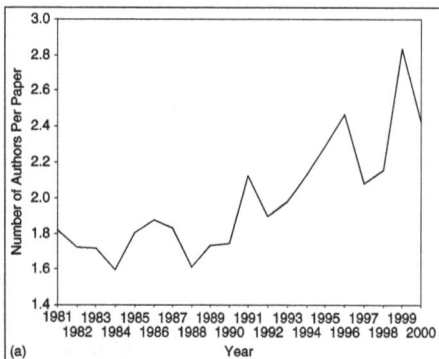

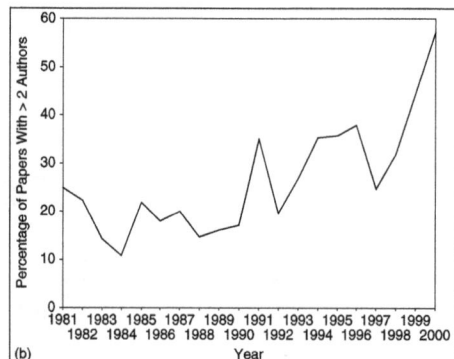

In figure, (a) Trend in the mean number of authors per paper published in the International Journal of Climatology (b) Trend in percentage of papers published in the International Journal of

Climatology with more than two authors from climatology has evolved rapidly as a credible and extremely important area of scientific activity. Its position today is the result of changing scientific, technological, social, economic, and political circumstances. These have created the opportunity for climatologists to not only conceptualise, observe and measure the nature of the climate system in different ways but also demonstrate that understanding the nature and causes of climate can produce significant benefits for coping with and managing climate as a determinant, hazard and resource. Consequently, an expansion of the scope of climatology beyond its traditional base of synoptic, dynamic, physical and statistical climatology, explicitly embracing areas such as climate and human affairs, is to be encouraged.

Climatic Geomorphology

Climatic geomorphology is the study of the role of climate in shaping landforms and the earth-surface processes. An approach used in climatic geomorphology is to study relict landforms to infer ancient climates. Being often concerned about past climates climatic geomorphology considered sometimes to be an aspect of historical geology. Since landscape features in one region might have evolved under climates different from those of the present, studying climatically disparate regions might help understand present-day landscapes. For example, Julius Büdel studied both cold-climate processes in Svalbard and weathering processes in tropical India to understand the origin of the relief of Central Europe, which he argued was a palimpsest of landforms formed at different times and under different climates.

Sub-disciplines

Desert Geomorphology

Desert geomorphology or the geomorphology of arid and semi-arid lands shares many landsforms and processes with more humid regions. One distinctive feature is the sparse or lacking vegetation cover, which influences fluvial and slope processes, related to wind and salt activity. Early work on desert geomorphology was done by Western explorers of the colonies of their respective countries in Africa (French West Africa, German South West Africa, Western Egypt), in frontier regions of their own countries (American West, Australian Outback) or in the deserts of foreign countries such as the Ottoman Empire, the Russian Empire and China. Since the 1970s desert geomorphology in Earth has served to find analogues to Martian landscapes.

Periglacial Geomorphology

As a discipline periglacial geomorphology is close but different to Quaternary science and geocryology. Periglacial geomorphology is concerned with non-glacial cold-climate landforms in areas with and without permafrost. Albeit the definition of what a periglacial zone is not clear-cut a conservative estimate is that a quarter of Earth's land surface has periglacial conditions. Beyond this quarter an additional quarter or fifth or Earth's land surface had periglacial conditions at some time during the Pleistocene. In periglacial geomorphology noted researchers include Johan Gunnar Andersson, Walery Łoziński, Anders Rapp and Jean Tricart.

Tropical Geomorphology

If the tropics is defined as the area between 35° N and 35° S, then about 60% of Earth's surface lies within this zone. During most of the 20th century tropical geomorphology was neglected due to a bias towards temperate climates, and when dealt with it was highlighted as "exotic". Tropical geomorphology do mainly differ from other areas in the intensities and rates at which surface processes operate, and not by the type of processes. The tropics are characterized by particular climates, that may be dry or humid. Relative to temperate zones the tropics contain areas of high temperatures, high rainfall intensities and high evapotranspiration all of which are climatic features relevant for surface processes. Another characteristic, that is not related to present-day climate, is that a large portion of the tropics have a low relief which was inherited from the continent of Gondwana. Julius Büdel, Pierre Birot and Jean Tricart have suggested that tropical rivers are dominated by fine-grained suspended load derived from advanced chemical weathering, and this would make them less erosive than rivers elsewhere. Some landforms previously thought as typically tropical like bornhardts are more related to lithology and rock structure than climate.

Morphoclimatic Zones

Yardangs in Lut Desert, Kerman Province, Iran. Deserts constitute undisputed morphoclimatic zones.

Climatic geomorphologists have devised various schemes that divide Earth's surface into various morphoclimatic zones; that is, zones where landforms are associated to present or past climates. However, only some processes and landforms can be associated with particular climates, meaning that they are *zonal;* processes and landforms not associated with particular climates are labelled *azonal.* Despite this, azonal processes and landforms might still take on particular characteristics when developing under the influence of particular climates. When identified, morphoclimatic zones do usually lack sharp boundaries and tend to grade from one type to another resulting in that only the core of the zone has all expected attributes. Influential morphoclimatic zoning schemes are those of Julius Büdel and of Jean Tricart and André Cailleux. Büdel's schemes stresses planation and valley-cutting in relation to climate, arguing the valley-cutting is dominant in sub-polar regions while planation is so in the tropics. As such this scheme is concerned not only with processes but also with end-products of geomorphic activity. The scheme of Tricart and Cailleux emphasizes the relationship between geomorphology, climate and vegetation. An early attempt at morphoclimatic zoning is that of Albrecht Penck in 1910, who divided Earth in three zones depending on the evaporation-precipitation ratios.

A 1994 review argues that only the concepts of desert, glacial, periglacial and a few coastal morpho-climatic zones are justified. These zones amounts to about half of Earth's land surface, the remaining half cannot be explained in simple terms by climate-landform interactions. The limitations of morphoclimatic zoning were already discussed by Siegfried Passarge in 1926 who considered vegetation and the extent of weathered material as having more direct impact than climate in many parts of the World. According to M.A. Summerfield large-scale zoning of the relief of Earth's surface is better explained on the basis of plate tectonics than on climate. An example of this are the Scandinavian Mountains whose plateau areas and valleys relate to the history of uplift and not to climate.

Climatic Elements

Weather Elements

The weather elements that are used to describe climate are also the elements that determine the type of climate for a region. The most important elements of weather which in different combinations make up the climate of a particular place or area are: solar radiation, air temperature, air pressure, wind velocity and wind direction, humidity and precipitation, and amount of cloudiness. The climatic elements of temperature, precipitation, and wind are the most significant elements used to express the climate of a region.

Temperature

The temperature of an area is dependent upon:

- Latitude or the distribution of the incoming and outgoing radiation,
- The nature of the surface (land or water),
- The altitude and the,
- Prevailing winds.

The air temperature normally used in climatology is that recorded at the surface. Moisture or lack of moisture modifies the temperature. The more moisture in a region, the smaller the temperature range, and the drier the region, the greater the temperature range. Moisture is also influenced by temperature. Warmer air can hold more moisture than a cooler air, resulting in increased evaporation and a higher probability of clouds and precipitation. Moisture, when coupled with condensation and evaporation, is an extremely important climatic element. It ultimately determines the type of climate for a specific region.

Precipitation

Precipitation is the second most important climatic element. In most studies, precipitation is defined as water reaching Earth's surface by falling either in a liquid or a solid state. The most significant forms are rain and snow. Precipitation has a wide range of variability over the earth's surface. Because of its variability, a longer series of observations is generally required to establish mean or an average. It often becomes necessary to include such factors as average number of days with precipitation and average amount per day. Precipitation is expressed in millimeters. Since precipitation amounts are directly associated with amounts and type of clouds, cloud cover must also be considered with precipitation. Cloud climatology also includes phenomena such as fog and thunderstorms.

Wind

Wind is the climatic element that transports heat and moisture into a region. Climatologists are mostly interested in wind with regards to its direction, speed, and gustiness. Wind is therefore usually discussed in terms of prevailing direction, average speeds, and maximum gusts. Some climatological studies use resultant wind, which is the vectorial average of all wind directions and speeds for a given level, at a specific place, and for a given period.

Expression of Climatic Elements

Climatic elements are observed over long periods of time; therefore, specific terms must be used to express these elements so they have a definite meaning.

Mean (Average)

The mean is the most commonly used climatological parameter. The term mean normally refers to a mathematical averaging obtained by adding the values of all factors or cases and then dividing by the number of items. For example, the average daily temperature would be the sum of the hourly temperatures divided by 24.

The mean temperature of 1 day has been devised by simply adding the maximum and minimum temperature values for that day and dividing by 2. In analyzing weather data, the terms average and mean are often used interchangeably.

Normal

In climatology, the term normal is applied to the average value over aperiod of time, which serves as a standard with which values (occurring on a date or during a specified time) may be compared. These periods of time may be a particular month or other portion of the year. They may refer to a season or to a year as a whole. The normal is usually determined over a 20- or 30-year period.

Absolute

In climatology, the term absolute is usually applied to the extreme highest and lowest values for any given meteorological element recorded at the place of observation and are most frequently applied to temperature.

Extreme

The term extreme is applied to the highest and lowest value for a particular meteorological element occurring over a period of time. This period of time is usually a matter of months, seasons or years.

The term may be used for a calendar day only, for which it is particularly applicable to temperature. At time the term is applied to the average of the highest and lowest temperatures as mean monthly or mean annual extremes.

Range

Range is the difference between the highest and lowest values and reflects the extreme variations of these values. Since it has a high variability, this statistic is not recommended for precise work. Range is related to extreme values of record and can be useful in determining the extreme range for the records available.

Frequency

Frequency is defined as the number of times a certain value occurs within a specified period of time. A frequency distribution may be used to present a condensed presentation of large data.

Mode

Mode is defined as the value occurring with the greatest frequency or the value about which the most cases occur.

Median

The median is the value at the midpoint in an array. It is determined by arranging all values in order of size. Rough estimates of the median may be obtained by taking the middle value of an ordered series; if there are two middle values they may be averaged to obtain the median. The median is not widely used in climatological computations. A longer period of record might be required to formulate an accurate median.

Degree-day

A degree-day is the number of degrees the mean daily temperature is above or below a standard temperature base. Degree days are accumulated over a season. At any point in the season, the total can be used as an index of past temperature effect upon some quantity, such as plant growth, fuel consumption, power output, etc. Degree-days are frequently applied to fuel and power consumption in the form of heating degree-days and cooling degree-days.

Average and Standard Deviations

In the analysis of climatological data, it may be desirable to compute the deviation of all items from a central point. This can be obtained from a computation of either the mean (or average) deviation or the standard deviation. These are termed measures of dispersion and are used to determine whether the average is truly representative or to determine the extent to which data vary from the average.

Average Deviation

Average deviation is obtained by computing the arithmetic average of the deviations from an average of the data.

Average Deviation = $\sum d / n$

Where d (the deviations) and n is the number of items.

Standard Deviation

Standard deviation is the measure of the scatter or spread of all values in a series of observations.

Standard deviation = $SQRT (\sum d^2 / n)$

Where d^2 is the sum of the squared deviations from the arithmetic average, and n is the number of items in the group of data.

Climatic Zones

The basic grouping of areas into climatic zones consists of classifying climates into five broad belts based on astronomical or mathematical factors. Actually they are zones of sunshine or solar climate and include the torrid or tropical zone, the two temperate zones, and the two polar zones.

The tropical zone is limited on the Tropic of Cancer and on the south by the Tropic of Cancer which are located at 23 1/2° north and south latitude, respectively.

The temperate zone of the Northern Hemisphere is limited on the south by the Tropic of Cancer and on the north by the Arctic Circle located at 66 1/2° north latitude. The temperate zone of the Southern Hemisphere is bounded on the north by the Tropic of Capricorn and on the south by the Antarctic Circle located at 66 1/2° south latitude.

The two polar zones are the areas in the Polar Regions which have the Arctic and Antarctic Circles as their boundaries.

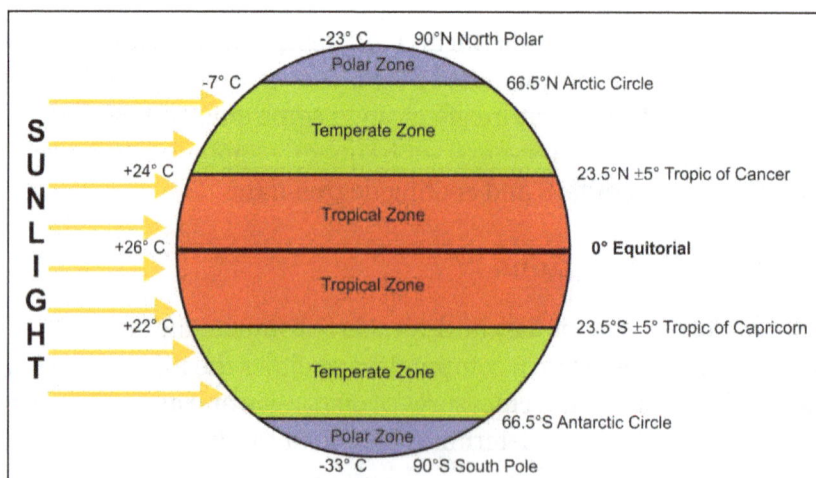

Technically, Climatic zones are limited by isotherms rather than by parallels of latitude. A glance at any chart depicting the isotherms over the surface of the earth shows the isotherms do not coincide with latitude lines. The astronomical or latitude zones therefore differ from the zones of heat.

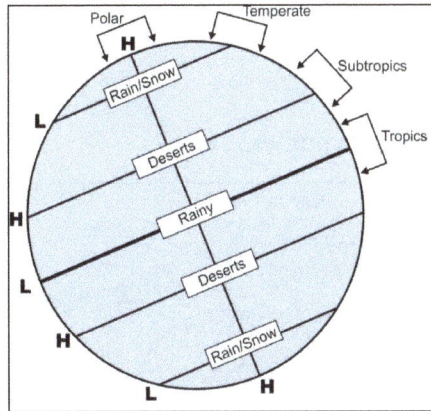

Electromagnetic energy travels in waves and spans a broad spectrum from very long radio waves to very short gamma rays. The human eye can only detect only a small portion of this spectrum called visible light. Our Sun is a source of energy across the full spectrum, and its electromagnetic radiation bombards our atmosphere constantly. However, the Earth's atmosphere protects us from exposure to a range of higher energy waves that can be harmful to life. Examples are Gamma rays, x-rays, and some ultraviolet waves. Electromagnetic radiation is reflected or absorbed mainly by several gases in the Earth's atmosphere, among the most important being water vapor, carbon dioxide, and ozone. Some radiation, such as visible light, largely passes (is transmitted) through the atmosphere. These regions of the spectrum with wavelengths that can pass through the atmosphere are referred to as "atmospheric windows." Some microwaves can even pass through clouds, which make them the best wavelength for transmitting satellite communication signals.

The Electromagnetic Spectrum

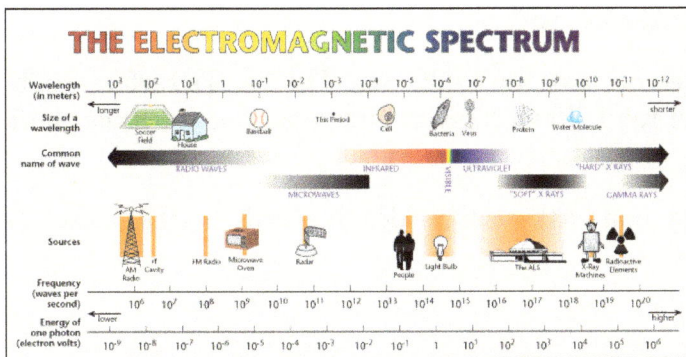

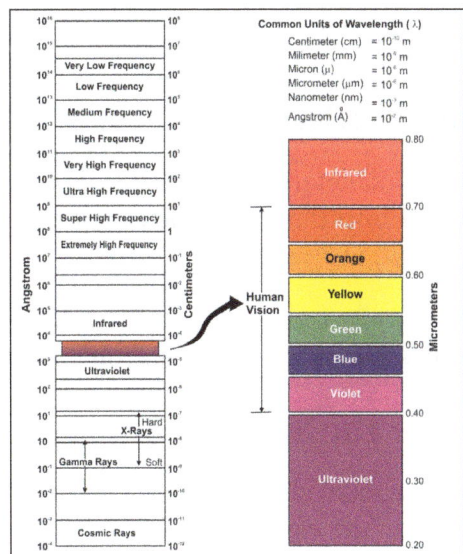

Eectromagnetic Spectrum.

The cause of seasons is due to orbit of the earth round the sun. This is shown below:

- Winter -December to February.

- Spring-March to May.

- Summer-June to August.

- Fall (autumn)-September to November.

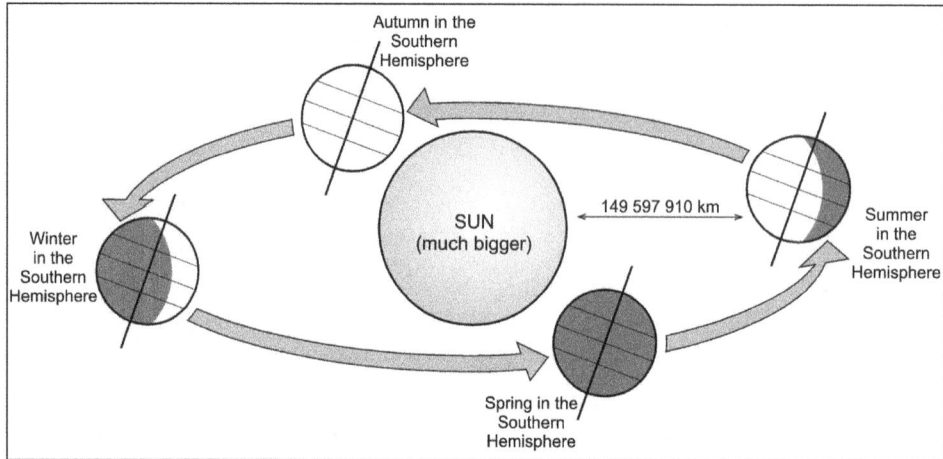

Orbit of the Earth round the Sun.

However, On the basis of climate, the period of year has been divided into four seasons in India. They are:

- Cold weather season (winter season) – January and February.

- Pre-monsoon or Hot weather season – March to May.

- SW or Summer Monsoon season – June to September.

- Post monsoon season – October to December.

Characteristics of the Earth's Climate Zone

Earth's planetary climate depends on its relative position to the Sun. The Earth's surface can be divided into three climatic zones based on rainfall and temperature controlled by atmospheric convection currents.

The Köppen-Geiger climate classification system further subdivides the Earth's surface based on rainfall, temperature and seasonal patterns.

Earth: The Habitable Planet

Earth's global climate consists of the averages of all the regional climates. The global climate depends on the energy received from the Sun and how much energy stays trapped in the planetary system. These factors change from planet to planet. The factors that make Earth tolerable for life (as we know life) start, like all good real estate, with location.

The Earth revolves around the Sun at a distance that keeps the overall temperature comfortable. In addition, the Earth sits at a distance that reduces the Sun's destructive radiation to a tolerable level.

The Earth consists of a rocky ball rather than a gaseous sphere. The Earth has a molten outer and solid inner iron-nickel core, however, which spins and generates a magnetic field.

The magnetic field helps deflect bursts of deadly solar radiation. The core also helps provide a source of geothermal heat to the mantle and, eventually, to the crust. The Earth also has an atmosphere. The current nitrogen-oxygen-argon atmosphere contains enough carbon dioxide and water vapor to trap the Sun's heat energy while also providing protection from radiation.

Earth's Major Climatic Zones

The Earth's surface can be divided into three major regional zones based on the three global convection cells that control average rainfall and average temperatures. The edges of the zones fall roughly along lines of latitude. The three zones are the tropical zone, the temperate zone and the polar zone. These zones have been subdivided using the Köppen-Geiger climate classification system.

Two Köppen-Geiger climate zones that occur across the three major regional zones are the Dry Zone and the Polar-Highland Subclimate. The Dry Zone is subdivided into the Desert Subclimate where average annual rainfall is less than 10 inches per year and the Semiarid Subclimate where rainfall averages slightly over 10 inches of rain per year.

In the Dry Zone, evaporation exceeds precipitation. The Dry Zone designation does not depend on temperature.

The Polar-Highland Subclimate has widely variable temperatures, depending on elevation, latitude and orientation. Elevation controls the climatic conditions in the Polar-Highland Subclimate. Mountains scattered around the world have Polar-Highland Subclimate conditions in their upper elevations.

Characteristics of the Tropical Zone

The tropical zone lies approximately between 25° north and 25° south latitudes. The tropical zone receives direct sunlight year-round, so the average temperature stays greater than 64 °F (18 °C) and the annual precipitation is greater than 59 inches. In the Köppen-Geiger climate classification system, the tropical zone is named the Humid Tropical Zone.

This zone has been subdivided into two subclimates, the Tropical Wet and the Tropical Wet & Dry. As the name indicates, the Tropical Wet Subclimate is hot and rainy all year. Tropical rain forests grow in this subclimate. The Tropical Wet & Dry Subclimate has distinct rainy and dry seasons.

Characteristics of the Temperate Zone

Characteristics of temperate climates are moderate temperatures and rain year-round. The local climates in the temperate zone show greater variability than the tropical zone, however. The temperate zone lies roughly between 25° and 60° north and south latitudes. At this point in geologic time, most of the Earth's land masses lie in the temperate zone.

In the Köppen-Geiger climate classification system, the temperate zone is divided into two zones: the Moist-mid Latitude - Mild Winters Zone and the Moist-mid Latitude - Severe Winters Zone.

The Moist-mid Latitude - Mild Winters Zone is subdivided into three subclimates: Humid Subtropical, Marine West Coast and Mediterranean.

As the name implies, these temperate areas share the characteristic of relatively mild weather, even in winter. The Moist-mid Latitude - Severe Winters Zone is subdivided into two subclimates: Humid Continental and Subarctic. Both subclimates experience cold snowy winters. The Humid Continental Subclimate has hot, humid summers while the Subarctic Subclimate endures short summers and long winters.

Characteristics of the Polar Zones

The polar zones extend from 60 °N and 60 °S latitudes to the north and south poles, respectively. In general, the variability of sunlight controls the climate characteristics of polar zones because each pole spends part of the year with no sunlight.

Even during each pole's summer, the sunlight hits at an angle that greatly reduces the heat energy. The annual temperatures for the polar zones almost always average below freezing with even the warmest month averaging below 50 °F (10 °C).

In the Köppen-Geiger climate classification system, the Polar Zone is subdivided into three Subclimates: Tundra, Icecap and Highland. The Tundra Subclimate typically is cold and dry with short cold summers. The Icecap Subclimate fits its name with freezing temperatures all year. The Highland Subclimate, occurs at higher elevations around the world.

References

- Climatology-IMTC, training: imdpune.gov.in, Retrieved 16 January, 2019

- Gutiérrez, Mateo, ed. (2005). "Chapter 1 Climatic geomorphology". Developments in Earth Surface Processes. 8. Pp. 3–32. Doi:10.1016/S0928-2025(05)80051-3. ISBN 978-0-444-51794-4

- Climatology-importance: omicsonline.org, Retrieved 29 March, 2019

- Migoń, Piotr (2006). "Büdel, J. 1982: Climatic geomorphology. Princeton: Princeton University Press. (Translation of Klima-geomorphologie, Berlin-Stuttgart: Gebrüder Borntraeger, 1977.)". Progress in Physical Geography. 30 (1): 99–103. Doi:10.1191/0309133306pp473xx

- Info-8521357-main-characteristics-earths-climate-zone: sciencing.com, Retrieved 2 June, 2019

Sub-divisions of Climatology

3

- **Paleoclimatology**
- **Paleotempestology**
- **Tropical Cyclone Rainfall Climatology**
- **Bioclimatology**
- **Historical Climatology**
- **Urban Climatology**
- **Dendroclimatology**
- **Synoptic Climatology**
- **Hydroclimatology**
- **Other Sub-divisions**

Climatology can be sub-divided into numerous fields such as paleoclimatology, paleotempestology, tropical cyclone rainfall climatology, bioclimatology, historical climatology, urban climatology, dendroclimatology, synoptic climatology, hydroclimatology, etc. The topics elaborated in this chapter will help in gaining a better perspective about these sub-fields of climatology.

Paleoclimatology

Paleoclimatology is the study of ancient climates, prior to the widespread availability of instrumental records. Similar to the way archeologists study fossils and other physical clues to gain insight into the prehistoric past, paleoclimatologists study several different types of environmental evidence to understand what the Earth's past climate was like and why. Over the years, the Earth has kept records of its climate conditions preserved in tree rings, locked in the skeletons of tropical coral reefs, sealed in glaciers and ice caps, and buried in laminated sediments from lakes and the ocean. Scientists can use those environmental recorders to estimate past conditions, extending our understanding of climate back hundreds to millions of years.

Trees like the Giant Sequoia pictured here can grow to be hundreds or even thousands of years old, providing an important source of information about past environmental variations.

If there is one thing that the paleoclimate record shows, it's that the Earth's climate is always changing. Over the past two million years, numerous glacial periods have covered much of the high-latitude northern hemisphere landmasses in glacial ice, dropped sea level as much as 410 feet, and significantly cooled even tropical regions. In the more distant past, the Cretaceous Period (between 145.5 and 65.5 million years ago) was significantly warmer than today, with less polar ice, rising sea levels, and warm weather organisms thriving even in near-polar regions.

The study of paleoclimates has been particularly helpful in showing that the Earth's climate system can shift between dramatically different climate states in a matter of years or decades. For example, tree-ring and lake-sediment records from North America show that decadal-scale "megadroughts" occurred multiple times over the last thousand years. During these periods, persistent droughts lasted much longer than any of the droughts we have experienced over the period of instrumental records. Understanding "climate surprises" of the past is critical to avoid being surprised by abrupt climatic change in the future.

The study of past climate change also helps us understand how humans influence the Earth's climate system. The climatic record over the last thousand years clearly shows that global temperatures increased significantly in the 20th century, and that this warming was likely to have been unprecedented in the last 1,200 years. The paleoclimatic record also allows us to examine the causes of past climate change and to help unravel how much of the 20th century warming may be explained by natural causes, such as solar variability, and how much may be explained by human influences.

Paleotempestology

Paleotempestology is the study of past tropical cyclone activity by means of geological proxies as well as historical documentary records. The term was coined by Kerry Emanuel.

Paleotempestology usually tries to identify leftover deposits from past storms, such as overwash deposits in waterbodies close to the coast which is the most commonly applied techniques, oxygen isotope ratio variations caused by tropical cyclone rainfall in trees or speleothems and beach ridges kicked up by storm waves. From these deposits one can then infer the occurrence rate of tropical cyclones - typically the stronger events are the most easily recognizable ones - and sometimes also their intensity, by comparing them to deposits left by historical events.

While the findings are prone to confounding factors and only some parts of the world have been investigated, some important findings have been made with the help of paleotempestology. For example, in the Gulf Coast and in Australia the occurrence rate of intense tropical cyclones is about once every few centuries, and there are long-term variations in occurrence which are caused e.g. by shifts in their paths.

Rationale

Tropical cyclones - known depending on location as cyclones, hurricanes or typhoons - with their precipitation, storm surges and winds are highly destructive and deadly phenomena; the 1900 Galveston hurricane claimed over 8,000 fatalities and was the worst natural disaster in United States history, while Hurricane Katrina in 2005 became the costliest hurricane in United States history with over 80 billion dollars damage and over 1600 fatalities. In other parts of the world, a 1970 cyclone killed 300,000 in Bangladesh; Japan in 2004 was hit by 10 typhoons and in 2005 five separate cyclones hit the Cook Islands in a short timeframe; a year later records were broken by Typhoon Saomai in China and Cyclone Larry in Australia. Finally, in 2013 Typhoon Haiyan became one of the most intense tropical cyclones ever recorded and caused 6,000 fatalities in the Philippines. Further, increased coastal development in general and in the United States rapid population growth along hurricane-prone coasts is creating additional attention to the danger posed by tropical cyclones and the interest in the hazard existing for major cities like Miami and New Orleans. Tropical cyclones can also have positive effects on society, for example by bringing rain to drought-prone regions. Finally, there is increasing evidence that tropical cyclone influence the climate themselves by enhancing poleward heat transport.

The historical record in many places is too short (one century at most) to properly determine the hazard produced by tropical cyclones, especially the rare very intense ones which at times are undersampled by historical records; in the United States for example only about 150 years of record are available and only a small number of hurricanes classified as category 4 or 5 - the most destructive ones in the Saffir-Simpson scale - have come ashore, making it difficult to estimate the hazard level, and elsewhere the record often goes back less than half a century. Such records may also not be representative for future climates. The realization that one cannot rely solely on historical records to infer past storm activity was a major driving force for the development of paleotempestology.

Information about past tropical cyclone occurrences can be used to constrain how their occurrences may change in the future or about how they respond to large-scale climate modes such as sea surface temperature changes. In general, the origin and behaviour of tropical cyclone systems is poorly understood and there is concern that man-made global warming will increase the intensity of tropical cyclones and the frequency of strong events by increasing sea surface temperatures.

Techniques

In general, paleotempestology is a complex field of science that overlaps with other disciplines like climatology and coastal geomorphology. A number of techniques have been used to estimate the past hazards from tropical cyclones. Many of these techniques have also been applied to studying extratropical storms, although research on this field is less advanced than on tropical cyclones.

Overwash Deposits

Overwash deposits in atolls, coastal lakes, marshes or reef flats are the most important paleoclimatological evidence of tropical cyclone strikes; when storms hit these areas currents and waves can overtop barriers, erode these and other beach structures and lay down deposits in the water bodies behind barriers. Isolated breaches and especially widespread overtopping of coastal barriers during storms can generate fan-like, layered deposits behind the barrier. Individual layers can be correlated to particular storms in favourable circumstances; in addition they are often separated by a clear boundary from earlier sediments. Such deposits have been observed in North Carolina after Hurricane Isabel in 2003, for example.

Several techniques have been applied to separate out storm overwash deposits from other sediments:

- Compared to the normal sedimentation processes in such places, tropical cyclone deposits are rougher and can be detected with sieving, laser-dependent technologies or x-ray fluorescence techniques.

- In sediment cores, deposits formed by tropical cyclones may be denser due to a larger proportion of mineral content associated with overwashes, which can be detected with x-ray fluorescence techniques.

- They may contain less organic matter than deposits formed through steady sedimentation, which can be detected by combusting the deposits and measuring the resulting mass loss. This and sediment grain sizes are the most common research tools for sediment cores.

- A little used technique is the analysis of organic material in sediment cores; there are characteristic changes in carbon and nitrogen isotope ratios after flooding and the entering of seawater, including a general increase in biological productivity.

- Overwash deposits can contain elements that do not normally occur at the site, such as strontium; this can be detected with x-ray fluorescence techniques.

- Finally, overwash deposits have usually brighter colours than those generated during steady sedimentation.

- Storm surges can transport living structures into such deposits that do not normally occur in these settings, although non-storm related salinity variations caused e.g. by droughts or non-storm related entry of water are a potential limitation of this method. Thus, this method is often supplemented with other proxies. The most common living structure employed here are foraminifera, although bivalves, diatoms, dinoflagellates, ostracods and pollen have also been used. Marine foraminifera however are not always present in deposits formed by historical storms.

The position of the research site vis-a-vis the storm path is an important factor in determining the usefulness of the record, which is also influenced by the geography of the analysis site such as e.g. vegetation cover. Prerequisites for successful correlation of overwash deposits to tropical cyclones are:

- The absence of tsunamis in the region, as their deposits can usually not be easily distinguished from storm deposits.

- The investigation area should have low biological activity, as bioturbation can otherwise erase evidence of storm deposits. Low biological activity can be found in sites with high salt or low oxygen concentrations.

- A high geomorphic stability of the site.

- High sedimentation rates can facilitate the preservation of storm deposits.

- Tides can destroy layered storm deposits; thus non-tidal waterbodies are ideally used. In tidally active waterbodies, correlations involving various sediment cores can be applied.

Dating and Intensity Determination

Various dating techniques can then be used to produce a chronology of tropical cyclone strikes at a given location and thus a recurrence rate; for example, at Lake Shelby in Alabama a return period of once every 318 years was determined for storms with windspeeds of over 54 metres per second (120 mph)-73 metres per second (160 mph).

The intensity and impacts of the tropical cyclone can also be inferred from overwash deposits by comparing the deposits to these formed by known storms and analyzing their lithology. Additionally, thicker sediment layers usually correspond to stronger storm systems. This procedure is not always clear-cut however.

For dating purposes radiometric dating procedures involving carbon-14, cesium-137 and lead-210 are most commonly used, often in combination, although uranium series dating, optically stimulated luminescence and correlations to humans land use can also be used in some places.

Beach Ridges

Beach ridges and cheniers form when storm surges, storm waves or tides deposit debris in ridges, with one ridge typically corresponding to one storm. Ridges can be formed by coral rubble where the coast coincides with coral reefs and can contain complicated layer structures, shells, pumice and gravel. A known example is the ridge that Cyclone Bebe generated on Funafuti atoll in 1971.

Beach ridges are common on the deltaic shores of China and are indicative of increased typhoon activity. They have also been found on the Australian coast facing the Great Barrier Reef and are formed from reworked corals. The height of each ridge appears to correlate with the intensity of the storm that produced it and thus the intensity of the forming storm can be inferred by numerical modelling and comparison to known storms and known storm surges. Ridges become older the farther inland they are; they can also be dated through optically stimulated luminescence and radiocarbon dating. In addition, no tsunami-generated beach ridges have been observed, and tsunamis are important confounding factors in paleotempestology.

However, wind-driven erosion or accumulation can alter the elevation of such ridges, and in addition the same ridge can be formed by more than one storm event as has been observed in Australia. Beach ridges can also shift around through non-storm processes after their formation and finally beach ridges can form through non-tropical cyclone processes. Sedimentary texture can be used to infer the origin of a ridge from storm surges.

Isotope Ratios

Precipitation in tropical cyclones has a characteristic isotope composition with a depletion of heavy oxygen isotopes; carbon and nitrogen isotope data have also been used to infer tropical cyclone activity. Corals can store oxygen isotope ratios which in turn reflect water temperatures, precipitation and evaporation; these in turn can be related to tropical cyclone activity. Fish otoliths and bivalves can also store such records, as can trees where the oxygen isotope ratios of precipitation are reflected in the cellulose of trees, and can be inferred with the help of tree rings. However, confounding factors like physiological isotope variations and soil properties also influence oxygen isotope ratios of tree cellulose, which can thus be only used to infer the frequency of storms and not their intensity.

Speleothems, deposits formed in caves through the dissolution and redeposition of dolomite and limestone, can store isotope signatures associated with tropical cyclones, especially in fast growing speleothems, areas with thin soils and speleothems which have undergone little alteration. Such deposits have a high temporal resolution and are also protected from many confounding factors although the extraction of annual layers has become possible only recently, with a two-week resolution (two separate layers correlated to two hurricanes that struck two weeks apart) achieved in one case. However, the suitability of speleothems depends on the characteristics of the cave they are found in; caves that flood frequently may have their speleothems eroded or otherwise damaged, for example, making them less suitable for paleotempestology research. Finally very old records can be obtained from oxygen isotope ratios in rocks.

Other Techniques

Historical documents such as county gazzettes in China, diaries, logbooks of travellers, official histories and old newspapers can contain information on tropical cyclones. In China such records go back over a millennium, while elsewhere it is usually confined to the last 130 years. Such historical records however are often ambiguous or unclear. The frequency of shipwrecks has been used to infer past tropical cyclone occurrence, such as has been done with a database of shipwrecks that the Spaniards suffered in the Caribbean.

Aside from oxygen isotope ratios, tree rings can also record information on storm-caused plant damage or vegetation changes such as thin tree rings due to storm-induced damage to a tree canopy and saltwater intrusion and the resulting slowdown in tree growth ("dendrotempestology"). Speleothems can also store trace elements which can signal tropical cyclone activity and mud layers formed by storm-induced cave flooding. Droughts on the other hand can cause groundwater levels to drop enough that subsequent storms cannot induce flooding and thus fail to leave a record, as has been noted in Yucatan.

Other techniques:

- Rhythmites in river mouths. These are formed when storms resuspend sediments; the sediments when the storm wanes fall out and form the deposits, especially in places with high sediment supplies. Carbon isotope and chemical data can be used to distinguish them from non-storm sedimentation.

- Sand dunes at coastlines is influenced by storm surge height, and sand splays can be formed when sand is swept off these dunes by storm surges and waves; such deposits however are better studied in the context of tsunamis and there is no clear way to distinguish between tsunami- and storm-formed splays.

- Hummocky deposits in shallow seas, known as tempestites. The mechanics of their formation are still controversial, and such deposits are prone to reworking which wipes out the traces of a storm.

- Boulders and coral blocks can be moved by storms and such moved blocks can potentially be dated to obtain the age of the storm, if certain conditions are met. They can be correlated to storms with the help of oxygen isotope excursions for example. This technique has also been applied to islands formed by storm-moved blocks.

- Wave-driven erosion during storms can create scarps which can be dated with the assistance of optically-stimulated luminescence. Such scarps however tend to be altered over time - later storms can erode away older scarps, for example - and their preservation and formation is often strongly dependent on the local geology.

- Other techniques involve the identification of freshwater flood deposits by storms such as humic acid and other evidence in corals, and lack of bromine - which is common in marine sediments - in flood-related deposits, and oyster bed kills caused by sediments suspended by storms (oyster kills however can also be caused by non-storm phenomena).

- Finally, luminescence of coral deposits has been used to infer tropical cyclone activity.

Timespans

A database going back to 8,000 BP has been compiled for the Atlantic Ocean. In the Gulf of Mexico, paleoproxy records go back five millennia but only a few typhoon records go back 5,000 - 6,000 years. In general, tropical cyclone records do not go farther back than 5,000 - 6,000 years ago when the Holocene sea level rise levelled off; tropical cyclone deposits formed during sea level lowstands likely were reworked during sea level rise. Only tentative evidence exists of deposits from the last interglacial. Tempestite deposits and oxygen isotope ratios in much older rocks have also been used to infer the existence of tropical cyclone activity as far back as the Jurassic.

Results

Paleotempestology is a field of science with important practical and social implications. The insurance industry factors in paleotempestological information in risk prediction analysis and when setting insurance rates, and it also funds paleotempestological research. Archeologists, ecologists, forest and water resource managers could also make use of paleotempestology information.

Recurrence Rates

The recurrence rate is an important metric with which one can estimate tropical cyclone risk, and it can be determined by paleotempestological research. In the Gulf of Mexico, catastrophic hurricane strikes at given locations occur once about every 350 years in the last 3,800 years or about 0.48%-0.39% annual frequency at any given site, with a recurrence rate of 300 years or 0.33% annual probability at sites in the Caribbean and Gulf of Mexico; category 3 or more storms occur at a rate of 3.9 - 0.1 category 3 or more storms per century in the northern Gulf of Mexico. Elsewhere, tropical cyclones with intensities of category 4 or more occur about every 350 years in the Pearl River Delta (China), 1 storm every 100–150 years at Funafuti and a similar rate in French Polynesia, 1 category 3 or stronger every 471 years in St. Catherines Island (Georgia), 1 storm every 140–180 years in Nicaragua, 1 intense storm every 200–300 years in the Great Barrier Reef - formerly their recurrence rate was estimated to be one strong event every few millennia - and 1 storm of category 2-4 intensity every 190–270 years at Shark Bay on the other side of Australia. Steady rates have been found for the Gulf of Mexico and the Coral Sea for timespans of several millennia.

However, it has also been found that the occurrence rates of tropical cyclone measured with instrumental data over historical time can be significantly different from the actual occurrent rate. In the past, tropical cyclones were far more frequent in the Great Barrier Reef and the northern Gulf of Mexico than today; in Apalachee Bay, strong storms occur every 40 years, not every 400 years as documented historically. and serious storms in New York occurred twice in 300 years not once every millennium or less. In general, the area of Australia appears to be unusually inactive in recent times by the standards of the past 550–1500 years, and that the historical record underestimates the incidence of strong storms in Northeastern Australia.

Long Term Fluctuations of Tropical Cyclone Activity

Long-term variations of tropical cyclone activity have also been found. The Gulf of Mexico saw increased activity between 3,800 - 1,000 years ago with a fivefold increase of category 4-5 hurricane activity, and activity at St. Catherines Island and Wassaw Island was also higher between 2,000 and 1,100 years ago. This appears to be a stage of increased tropical cyclone activity spanning the region from New York to Puerto Rico, followed by an inactive interval since 1,000 years that also affected the Gulf Coast. The US Atlantic coast and the Caribbean saw low activity between 950 AD and 1700 with a sudden increase around 1700. Such fluctuations appear to mainly concern strong tropical cyclone systems, at least in the Atlantic; weaker systems have a more steady pattern of activity. Rapid fluctuations over short timespans have also been observed.

In the Atlantic Ocean, the so-called "Bermuda High" hypothesis stipulates that changes in the position of this anticyclone can cause storm paths to alternate between landfalls on the East Coast and the Gulf Coast but also Nicaragua. Paleotempestological data support this theory although additional findings on Long Island and Puerto Rico have demonstrated some complexity in the patterns as active periods appear to correlate between the three sites. A southward shift of the High has been inferred to have occurred 3,000-1,000 years ago and has been linked with the "hurricane hyperactivity" period in the Gulf of Mexico between 3,400 - 1,000 years ago. Furthermore, a tendency to a more northerly storm track may be associated with a strong North Atlantic Oscillation while the Neoglacial cooling is associated with a southward shift. A north-south anti-correlation has also been found in West Asia between the South China Sea and Japan.

Influence of Climate Modes on Tropical Cyclone Activity

The influence of natural trends on tropical cyclone activity has been recognized in paleotempe-stology records, such as correlation of Atlantic hurricane tracks and activity with the status of the ITCZ, position of the Loop Current (for Gulf of Mexico hurricanes), North Atlantic Oscillation, sea surface temperatures and the strength of the West African Monsoon, and correlation between Australian cyclone activity and the Pacific Decadal Oscillation. Increased insolation - either from solar activity or from orbital variations - have been found to be detrimental to tropical cyclone activity in some regions. In the early Common Era, warmer sea surface temperatures in the Atlantic as well as more restricted anomalies may be responsible for stronger regional hurricane activity.

Among the known climate modes that influence tropical cyclone activity in paleotempestological records are ENSO phase variations, which influence tropical cyclone activity in Australia and the Atlantic but also their path as has been noted for typhoons. More general global correlations have been found, such as anticorrelation between tropical cyclone activity in Japan and the North Atlantic and correlation between the Atlantic and Australia on the one hand and between Australia and French Polynesia on the other hand.

Influence of Long-term Temperature Variations on Tropical Cyclone Activity

The effect of general climate variations have also been found. Hurricane and typhoon tracks tend to shift north (e.g. Amur Bay) during warm periods and south (e.g. South China) during cold periods, patterns that might be mediated by shifts in the subtropical anticyclones. Such a behaviour (northward shift) has been observed as a consequence of man-made global warming and the end of the Little Ice Age but also after volcanic eruptions (southward shift).

During the last 600 years in the Little Ice Age, there were more but weaker storms in the Gulf of Mexico while hurricane activity did not decrease in western Long Island. Increased hurricane activity during the last 300 years in the Caribbean may also correlate to the Little Ice Age.

The response of tropical cyclone to future global warming is of great interest. The Holocene Climatic Optimum did not induce increased tropical cyclone strikes in Queensland and phases of higher hurricane activity on the Gulf Coast are not associated with global warming; however warming has been correlated with typhoon activity in the Gulf of Thailand and marine warming with typhoon activity in the South China Sea, increased hurricane activity in Belize (which increased during the Medieval Warm Period) and during the Mesozoic, when carbon dioxide caused warming episodes.

After-effects of Tropical Cyclone Activity

A correlation between hurricane strikes and subsequent wildfire activity and vegetation changes has been noted in Alabama and Cuba paleotempestological evidence. In St. Catherines Island cultural activity ceased at the time of increased storm activity, and both Taino settlement in the Bahamas and Polynesian expansion across the Pacific may have been correlated to decreased tropical cyclone activity. Finally, tropical cyclone induced alteration in oxygen isotope ratios may mask isotope ratio variations caused by other climate phenomena, which may thus be misinterpreted.

On the other hand, the Classic Maya collapse may coincide with, and have been caused by, a decrease in tropical cyclone activity, and tropical cyclones are more important for drought amelioration in the southeastern USA than was appreciated.

Problems

Not all of the world has been investigated with paleotempestological methods; among the places thus researched are Belize, the Carolinas of North America, northern coasts of the Gulf of Mexico, the northeastern United States, (in a lesser measure) the South Pacific islands and tropical Australia. Conversely China, Cuba, Florida, Hispaniola, Honduras, the Lesser Antilles and North America north of Canada are little researched. The frequency of research institutions active in paleotempestology, suitable sites for paleotempestological research and of tropical cyclone landfalls may influence whether a given location is researched or not. In the Atlantic Ocean research has been concentrated on regions where hurricanes are common rather than more marginal areas.

Paleotempestological reconstructions are subject to a number of limitations, including changes in the hydrological properties of the site due to e.g. sea level rise which increases the sensitivity to weaker storms and "false positives" caused by for example non-tropical cyclone-related floods, sediment winnowing, wind-driven transport, tides, tsunamis, bioturbation and non-tropical storms such as nor'easters or winter storm, the latter of which however usually result in lower surges. In particular, tsunamis are a problem for paleotempestological studies in the Indian and Pacific Ocean; one technique that has been used to differentiate the two is the identification of traces of runoff which occurs during storms but not during tsunamis.

Paleotempestology records mostly record activity during the Holocene and tend to record mainly catastrophic storms as these are the ones most likely to leave evidence. In addition, as of 2017 there has been little effort in making comprehensive databases of paleotempestological data or in attempting regional reconstructions from local results.

Additional problems which have been noted is the fact that paleotempestology records, especially overwash records in marshes, are often short and their geochronology questionable; the often poor documentation of the deposition mechanisms and of which microfossils can be used as proxies for storms. The magnitude of overwash deposits is fundamentally a function of storm surge height, which are not simply a function of storm intensity. Overwash deposits are regulated by the height of the overwashed barrier and there is no expectation that it will remain stable over time; tropical cyclones themselves have been observed eroding such barriers and such barrier height decreases (e.g. through storm erosion or sea level rise) may induce a spurious increase of tropical cyclone deposits over time. Successive overwash deposits can be difficult to distinguish, and they are easily eroded by subsequent storms. Finally, storm deposits can vary strongly even a short distance from the landfall point, and over few tens of metres.

Tropical Cyclone Rainfall Climatology

A tropical cyclone rainfall climatology is developed to determine rainfall characteristics of past tropical cyclones. A tropical cyclone rainfall climatology can be used to help forecast current or upcoming tropical cyclone impacts. The degree of a tropical cyclone rainfall impact depends upon speed of movement, storm size, and degree of vertical wind shear. One of the most significant threats from tropical cyclones is heavy rainfall. Large, slow moving, and non-sheared tropical cyclones produce the heaviest rains. The intensity of a tropical cyclone appears to have little bearing on its potential for rainfall over land, but satellite measurements over the last several years show that more intense tropical cyclones produce noticeably more rainfall over water. Flooding from tropical cyclones remains a significant cause of fatalities, particularly in low-lying areas.

Anticipating a Flood Event

While inland flooding is common to tropical cyclones, there are factors which lead to excessive rainfall from tropical cyclones. Slow motion, as was seen during Hurricane Danny and Hurricane Wilma, can lead to high amounts of rainfall. The presence of mountains/hills near the coast, like across much of Mexico, Haiti, the Dominican Republic, Central America, Madagascar, Réunion, China, and Japan acts to magnify rainfall potential due to forced upslope flow into the mountains. Strong upper level forcing from a trough moving through the Westerlies and its associated cold front, as was the case during Hurricane Floyd, can lead to high amounts even from systems moving at an average forward motion. Larger tropical cyclones drop more rainfall as they precipitate upon one spot for a longer time frame than average or small tropical cyclones. A combination of two of these factors could be especially crippling, as was seen during Hurricane Mitch in Central America. During the 2005 season, flooding related to slow-moving Hurricane Stan's broad circulation led to 1,662–2,000 deaths.

General Distribution within a Tropical Cyclone

Isaac Cline was the first to investigate rainfall distribution around tropical cyclones in the early 1900s. He found that a larger proportion of rainfall falls in advance of the center (or eye) than after the center's passage, with the highest percentage falling in the right front quadrant. Father Viñes of Cuba found that some tropical cyclones have their highest rainfall rates in the rear quadrant within a training (non-moving) inflow band. Normally, as a tropical cyclone intensifies, its heavier rainfall rates become more concentrated around its center. Rainfall is found to be heaviest in tropical cyclone's inner core, whether it be the eyewall or central dense overcast, within a degree latitude of the center, with lesser amounts farther away from the center. Most of the rainfall in tropical cyclones is concentrated within its radius of gale-force (34 knots/39 mph/63 km/h) winds. Rainfall is more common near the center of tropical cyclones overnight. Over land, outer bands are more active during the heating of the day, which can act to restrict inflow into the center of the cyclone. Recent studies have shown that half of the rainfall within a tropical cyclone is stratiform in nature. The chart to the right was developed by Riehl in 1954 using meteorological equations that assume a gale radius of about 140 miles (230 km), a fairly symmetric cyclone, and does not consider topographic effects or vertical wind shear. Local amounts can exceed this chart by a factor of two due to topography. Wind shear tends to lessen the amounts below what is shown on the table.

Rainfall Rate per day within radius of the center (Riehl)			
Radius (mi)	Radius (km)	Amount (in)	Amount (mm)
35	56	33.98	863
70	112	13.27	337
140	224	4.25	108
280	448	1.18	30

Relation to Storm Size

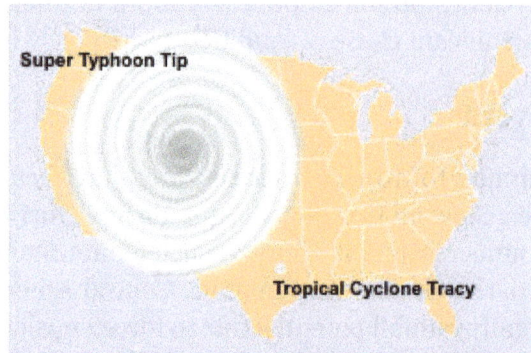

The relative sizes of Typhoon Tip, Cyclone Tracy.

Larger tropical cyclones have larger rain shields, which can lead to higher rainfall amounts farther from the cyclone's center. This is generally due to the longer time frame rainfall falls at any one spot in a larger system, when compared to a smaller system. Some of the difference seen concerning rainfall between larger and small storms could be the increased sampling of rainfall within a larger tropical cyclone when compared to that of a compact cyclone; in other words, the difference could be the result of a statistical problem.

Slow/looping Motion on Rainfall Magnitude

Storms which have moved slowly, or loop, over a succession of days lead to the highest rainfall amounts for several countries. Riehl calculated that 33.97 inches (863 mm) of rainfall per day can be expected within one-half degree, or 35 miles (56 km), of the center of a mature tropical cyclone. Many tropical cyclones progress at a forward motion of 10 knots, which would limit the duration of this excessive rainfall to around one-quarter of a day, which would yield about 8.50 inches (216 mm) of rainfall. This would be true over water, within 100 miles (160 km) of the coastline, and outside topographic features. As a cyclone moves farther inland and is cut off from its supply of warmth and moisture (the ocean), rainfall amounts from tropical cyclones and their remains decrease quickly.

Vertical Wind Shear Impact on Rainfall Shield

Vertical wind shear forces the rainfall pattern around a tropical cyclone to become highly asymmetric, with most of the precipitation falling to the left and downwind of the shear vector, or downshear left. In other words, southwesterly shear forces the bulk of the rainfall north-northeast of the center. If the wind shear is strong enough, the bulk of the rainfall will move away from the center leading to what is known as an exposed circulation center. When this occurs, the potential magnitude of rainfall with the tropical cyclone will be significantly reduced.

Effect of Interaction with Frontal Boundaries/Upper Level Troughs

As a tropical cyclone interacts with an upper-level trough and the related surface front, a distinct northern area of precipitation is seen along the front ahead of the axis of the upper level trough. This type of interaction can lead to the appearance of the heaviest rainfall falling along and to the left of the tropical cyclone track, with the precipitation streaking hundreds of miles or kilometers downwind from the tropical cyclone. The stronger the upper trough picking up the tropical cyclone, the more significant the left of track shift in the rainfall distribution tends to be.

Mountains

Moist air forced up the slopes of coastal hills and mountain chains can lead to much heavier rainfall than in the coastal plain. This heavy rainfall can lead to landslides, which still cause significant loss of life such as seen during Hurricane Mitch in Central America.

Global Distribution

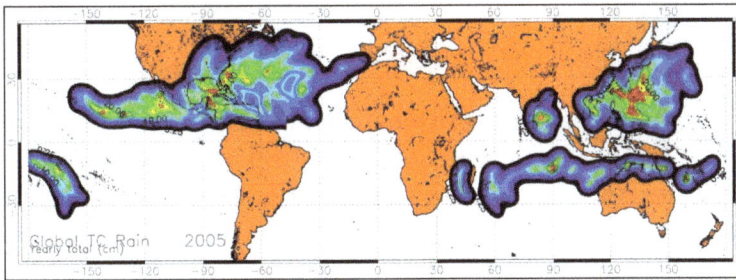

Global tropical cyclone rainfall.

Globally, tropical cyclone rainfall is more common across the northern hemisphere than across the southern hemisphere. This is mainly due to the normal annual tropical cyclone distribution, as between half and two-thirds of all tropical cyclones form north of the equator. Rainfall is concentrated near the 15th parallel in both hemispheres, with a less steep dropoff seen with latitude across the northern hemisphere, due to the stronger warm water currents seen in that hemisphere which allow tropical cyclones to remain tropical in nature at higher latitudes than south of the equator. In the southern hemisphere, rainfall impacts will be most common between January and March, while north of the equator, tropical cyclone rainfall impacts are more common between June and November. Japan receives over half of its rainfall from typhoons.

Bioclimatology

Bioclimatology is branch of climatology that deals with the effects of the physical environment on living organisms over an extended period of time. Although Hippocrates touched on these matters 2,000 years ago in his treatise on Air, Waters, and Places, the science of bioclimatology is relatively new. It developed into a significant field of study during the 1960s owing largely to a growing concern over the deteriorating environment.

Because almost every aspect of climate and weather has some effect on living organisms, the scope of bioclimatology is almost limitless. Certain areas are emphasized more than others, however, among them studies of the influence of weather and climate on small plant organisms and insects responsible for the development of plant, animal, and human diseases; the influence of weather and climate on physiological processes in normal healthy humans and their diseases; the influence of microclimate in dwellings and urban centres on human health; and the influence of past climatic conditions on the development and distribution of plants, animals, and humans.

Historical Climatology

Historical climatology is the study of historical changes in climate and their effect on human history and development. This differs from paleoclimatology which encompasses climate change over the entire history of Earth. The study seeks to define periods in human history where temperature or precipitation varied from what is observed in the present day. The primary sources include written records such as sagas, chronicles, maps and local history literature as well as pictorial representations such as paintings, drawings and even rock art. The archaeological record is equally important in establishing evidence of settlement, water and land usage.

Techniques

In literate societies, historians may find written evidence of climatic variations over hundreds or thousands of years, such as phenological records of natural processes, for example viticultural records of grape harvest dates. In preliterate or non-literate societies, researchers must rely on other techniques to find evidence of historical climate differences.

Past population levels and habitable ranges of humans or plants and animals may be used to find evidence of past differences in climate for the region. Palynology, the study of pollens, can show not only the range of plants and to reconstruct possible ecology, but to estimate the amount of precipitation in a given time period, based on the abundance of pollen in that layer of sediment or ice.

Evidence of Climatic Variations

The eruption of the Toba supervolcano, 70,000 to 75,000 years ago reduced the average global temperature by 5 degrees Celsius for several years and may have triggered an ice age. It has been postulated that this created a bottleneck in human evolution. A much smaller but similar effect occurred after the eruption of Krakatoa in 1883, when global temperatures fell for about 5 years in a row.

Before the retreat of glaciers at the start of the Holocene (~9600 BC), ice sheets covered much of the northern latitudes and sea levels were much lower than they are today. The start of our present interglacial period appears to have helped spur the development of human civilization.

Human Record

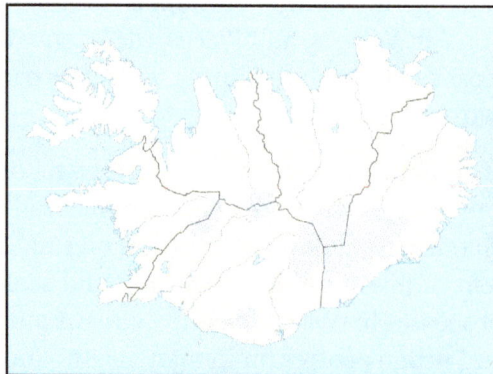

The Skálhol Map.

One of Grimspound's hut circles.

Evidence of a warm climate in Europe, for example, comes from archaeological studies of settlement and farming in the Early Bronze Age at altitudes now beyond cultivation, such as Dartmoor, Exmoor, the Lake district and the Pennines in Great Britain. The climate appears to have deteriorated towards the Late Bronze Age however. Settlements and field boundaries have been found at high altitude in these areas, which are now wild and uninhabitable. Grimspound on Dartmoor is well preserved and shows the standing remains of an extensive settlement in a now inhospitable environment.

Some parts of the present Saharan desert may have been populated when the climate was cooler and wetter, judging by cave art and other signs of settlement in Prehistoric Central North Africa.

The Medieval Warm Period was a time of warm weather between about AD 800–1300, during the European Medieval period. Archaeological evidence supports studies of the Norse sagas which describe the settlement of Greenland in the 9th century AD of land now quite unsuitable for cultivation. For example, excavations at one settlement site have shown the presence of birch trees during the early Viking period. The same period records the discovery of an area called Vinland, probably in North America, which may also have been warmer than at present, judging by the alleged presence of grape vines. The interlude is known as the Medieval Warm Period.

Little Ice Age

Later examples include the Little Ice Age, well documented by paintings, documents (such as diaries) and events such as the River Thames frost fairs held on frozen lakes and rivers in the 17th and 18th centuries. The River Thames was made more narrow and flowed faster after old London Bridge was demolished in 1831, and the river was embanked in stages during the 19th century, both of which made the river less liable to freezing. Among the earliest references to the coming climate change is an entry in the *Anglo-Saxon Chronicle* dated 1046:

> "And in this same year after the 2nd of February came the severe winter with frost and snow, and with all kinds of bad weather, so that there was no man alive who could remember so severe a winter as that, both through mortality of men and disease of cattle; both birds and fishes perished through the great cold and hunger."

The *Chronicle* is the single most important historical source for the period in England between the departure of the Romans and the decades following the Norman Conquest. Much of the information given in the *Chronicle* is not recorded elsewhere.

The Frozen Thames.

The Little Ice Age brought colder winters to parts of Europe and North America. In the mid-17th century, glaciers in the Swiss Alps advanced, gradually engulfing farms and crushing entire villages. The River Thames and the canals and rivers of the Netherlands often froze over during the winter, and people skated and even held frost fairs on the ice. The first Thames frost fair was in 1607; the last in 1814, although changes to the bridges and the addition of an embankment affected the river flow and depth, diminishing the possibility of freezes. The freeze of the Golden Horn and the southern section of the Bosphorus took place in 1622. In 1658, a Swedish army marched across the Great Belt to Denmark to invade Copenhagen. The Baltic Sea froze over, enabling sledge rides from Poland to Sweden, with seasonal inns built on the way. The winter of 1794/1795 was particularly harsh when the French invasion army under Pichegru could march on the frozen rivers of the Netherlands, while the Dutch fleet was fixed in the ice in Den Helder harbour. In the winter of 1780, New York Harbour froze, allowing people to walk from Manhattan to Staten Island. Sea ice surrounding Iceland extended for miles in every direction, closing that island's harbours to shipping.

The last written records of the Norse Greenlanders are from a 1408 marriage in Hvalsey Church — today the best-preserved of the Norse ruins.

The severe winters affected human life in ways large and small. The population of Iceland fell by half, but this was perhaps also due to fluorosis caused by the eruption of the volcano Laki in 1783. Iceland also suffered failures of cereal crops and people moved away from a grain-based diet. The Norse colonies in Greenland starved and vanished (by the 15th century) as crops failed and livestock could not be maintained through increasingly harsh winters, though Jared Diamond noted that they had exceeded the agricultural carrying capacity before then. In North America, American Indians formed leagues in response to food shortages. In Southern Europe, in Portugal, snow storms were much more frequent while today they are rare. There are reports of heavy snowfalls in the winters of 1665, 1744 and 1886.

In contrast to its uncertain beginning, there is a consensus that the Little Ice Age ended in the mid-19th century.

Evidence of Anthropogenic Climate Change

Through deforestation and agriculture, some scientists have proposed a human component in some historical climatic changes. Human-started fires have been implicated in the transformation of much of Australia from grassland to desert. If true, this would show that non-industrialized societies could have a role in influencing regional climate. Deforestation, desertification and the salinization of soils may have contributed to or caused other climatic changes throughout human history.

Urban Climatology

Urban climatology refers to a specific branch of climatology that is concerned with interactions between urban areas and the atmosphere, the effects they have on one another, and the varying spatial and temporal scales at which these processes (and responses) occur.

Luke Howard is considered to have established urban climatology with his book *The Climate of London*, which contained continuous daily observations from 1801 to 1841 of wind direction, atmospheric pressure, maximum temperature, and rainfall.

Urban climatology came about as a methodology for studying the results of industrialization and urbanization. Constructing cities changes the physical environment and alters energy, moisture, and motion regimes near the surface. Most of these alterations can be traced to causal factors such as air pollution; anthropogenic sources of heat; surface waterproofing; thermal properties of the surface materials; and morphology of the surface and its specific three-dimensional geometry—building spacing, height, orientation, vegetative layering, and the overall dimensions and geography of these elements. Other factors are relief, proximity to water bodies, size of the city, population density, and land-use distributions.

Influential Factors

Several factors influence the urban climate, including city size, the morphology of the city, land-use configuration, and the geographic setting (such as relief, elevation, and regional climate). Some of the differences between urban and rural climates include air quality, wind patterns, and changes in rainfall patterns, but one of the most studied is the urban heat island effect.

Temperature and Urban Heat Island Effect

Urban environments are typically warmer than their surroundings, as documented over a century ago by Howard. Urban areas are islands or spots on the broader scale compared with more rural surrounding land. The spatial distribution of temperatures occurs in tandem with temporal changes, which are both causally related to anthropogenic sources.

The urban environment has two atmosphere layers, besides the planetary boundary layer outside and extending well above the city: (1) The urban boundary layer is due to the spatially integrated heat and moisture exchanges between the city and its overlying air. (2) The surface of the city corresponds to the level of the urban canopy layer. Fluxes across this plane comprise those from individual units, such as roofs, canyon tops, trees, lawns, and roads, integrated over larger land-use divisions (for example, suburbs). The urban heat island effect has been a major focus of urban climatological studies, and in general the effect the urban environment has on local meteorological conditions.

Pollution

The field also includes the topics of air quality, Radiation Fluxes, Micro-Climates and even issues traditionally associated with architectural design and engineering, such as Wind Engineering. Causes and effects of pollution as understood through Urban Climatology are becoming more important for Urban Planning.

Precipitation

Changes in winds and convection patterns over and around cities impacts precipitation. Contributing factors are believed to be urban heat island, heightened surface roughness, and increased aerosol concentration.

Climate Change

Urban climatology is strongly linked to research surrounding global warming. As centers for socio-economic activities, cities produce large amounts of greenhouse gases, most notably CO_2 as a consequence of human activities such as transport, development, waste related to heating and cooling requirements etc.

Globally, cities are expected to grow into the 21st century (and beyond) - as they grow and develop the landscapes in which they inhabit will change so too will the atmosphere resting above them, increasing emissions of GHG's thus contributing to the global green house effect.

Finally, many cities are vulnerable to the projected consequences of climate change (sea level rise, changes in temperature, precipitation, storm frequency) as most develop on or near coast-lines, nearly all produce distinct urban heat islands and atmospheric pollution: as areas in which there is concentrated human habitation these effects potentially will have the largest and most dramatic impact and thus are a major focus for urban climatology.

Spatial Planning and Public Health

Urban Climatology impacts decision-making for municipal planning and policy in regards to pollution, extreme heat events, and stormwater modeling.

Dendroclimatology

Dendroclimatology is the science of determining past climates from trees (primarily properties of the annual tree rings). Tree rings are wider when conditions favor growth, narrower when times are difficult. Other properties of the annual rings, such as maximum latewood density (MXD) have been shown to be better proxies than simple ring width. Using tree rings, scientists have estimated many local climates for hundreds to thousands of years previous. By combining multiple tree-ring studies (sometimes with other climate proxy records), scientists have estimated past regional and global climates.

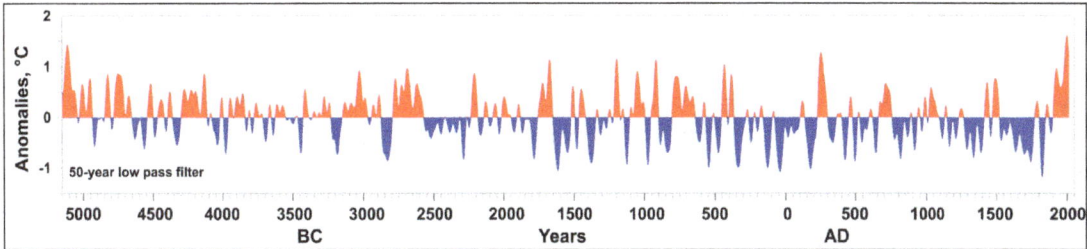

Variation of tree ring width translated into summer temperature anomalies for the past 7000 years, based on samples from holocene deposits on Yamal Peninsula and Siberian now living conifers.

Advantages

Tree rings are especially useful as climate proxies in that they can be well-dated via dendrochronology, i.e. matching of the rings from sample to sample. This allows extension backwards in time using deceased tree samples, even using samples from buildings or from archeological digs. Another advantage of tree rings is that they are clearly demarked in annual increments, as opposed to other proxy methods such as boreholes. Furthermore, tree rings respond to multiple climatic effects (temperature, moisture, cloudiness), so that various aspects of climate (not just temperature) can be studied. However, this can be a double-edged sword.

Limitations

Along with the advantages of dendroclimatology are some limitations: confounding factors, geographic coverage, annular resolution, and collection difficulties. The field has developed various methods to partially adjust for these challenges.

Confounding Factors

There are multiple climate and non-climate factors as well as nonlinear effects that impact tree ring width. Methods to isolate single factors (of interest) include botanical studies to calibrate growth influences and sampling of "limiting stands" (those expected to respond mostly to the variable of interest).

Climate Factors

Climate factors that affect trees include temperature, precipitation, sunlight, and wind. To differentiate among these factors, scientists collect information from "limiting stands." An example of a limiting stand is the upper elevation treeline: here, trees are expected to be more affected by temperature variation (which is "limited") than precipitation variation (which is in excess). Conversely, lower elevation treelines are expected to be more affected by precipitation changes than temperature variation. This is not a perfect work-around as multiple factors still impact trees even at the "limiting stand," but it helps. In theory, collection of samples from nearby limiting stands of different types (e.g. upper and lower treelines on the same mountain) should allow mathematical solution for multiple climate factors.

Non-climate Factors

Non-climate factors include soil, tree age, fire, tree-to-tree competition, genetic differences, logging or other human disturbance, herbivore impact (particularly sheep grazing), pest outbreaks, disease, and CO_2 concentration. For factors which vary randomly over space (tree to tree or stand to stand), the best solution is to collect sufficient data (more samples) to compensate for confounding noise. Tree age is corrected for with various statistical methods: either fitting spline curves to the overall tree record or using similar aged trees for comparison over different periods (regional curve standardization). Careful examination and site selection helps to limit some confounding effects, for example picking sites undisturbed by modern man.

Non-linear Effects

In general, climatologists assume a linear dependence of ring width on the variable of interest (e.g. moisture). However, if the variable changes enough, response may level off or even turn opposite. The home gardener knows that one can underwater or overwater a house plant. In addition, it is possible that interaction effects may occur (for example "temperature times precipitation" may affect growth as well as temperature and precipitation on their own. Here, also, the "limiting stand" helps somewhat to isolate the variable of interest. For instance, at the upper treeline, where the tree is "cold limited", it's unlikely that nonlinear effects of high temperature ("inverted quadratic") will have numerically significant impact on ring width over the course of a growing season.

Botanical Inferences to Correct for Confounding Factors

Botanical studies can help to estimate the impact of confounding variables and in some cases guide corrections for them. These experiments may be either ones where growth variables are all controlled (e.g. in a greenhouse), partially controlled (e.g. FACE [Free Airborne Concentration Enhancement] experiments—add ref), or where conditions in nature are monitored. In any case, the important thing is that multiple growth factors are carefully recorded to determine what impacts growth. (Insert Fennoscandanavia paper reference). With this information, ring width response can be more accurately understood and inferences from historic (unmonitored) tree rings become more certain. In concept, this is like the limiting stand principle, but it is more quantitative—like a calibration.

Divergence Problem

The divergence problem is the disagreement between the temperatures measured by the thermometers (instrumental temperatures) on one side, and the temperatures reconstructed from the latewood density or width of tree rings on the other side, at many treeline sites in northern forests.

While the rendering and analysis of data from thermometer records largely suggest a substantial warming trend, tree rings from these particular sites do not display a corresponding change in their maximum latewood density or, in some cases, their width. This does not apply to all such studies. Where this applies, a temperature trend extracted from tree rings alone would not show any substantial warming. The temperature graphs calculated from instrumental temperatures and from these tree ring proxies thus "diverge" from one another since the 1950s, which is the origin of the term. This divergence raises obvious questions of whether other, unrecognized divergences have occurred in the past, prior to the era of thermometers. There is evidence suggesting that the divergence is caused by human activities, and so confined to the recent past, but use of affected proxies can lead to overestimation of past temperatures, understating the current warming trend. There is continuing research into explanations and ways to reconcile this the discrepancy between analysis of tree ring data and thermometer based data.

Geographic Coverage

Trees do not cover the Earth. Polar and marine climates cannot be estimated from tree rings. In perhumid tropical regions, Australia and southern Africa, trees generally grow all year round and don't show clear annual rings. In some forest areas, the tree growth is too much influenced by multiple factors (no "limiting stand") to allow clear climate reconstruction. The coverage difficulty is dealt with by acknowledging it and by using other proxies (e.g. ice cores, corals) in difficult areas. In some cases it can be shown that the parameter of interest (temperature, precipitation, etc.) varies similarly from area to area, for example by looking at patterns in the instrumental record. Then one is justified in extending the dendroclimatology inferences to areas where no suitable tree ring samples are obtainable.

Annular Resolution

Tree rings show the impact on growth over an entire growing season. Climate changes deep in the dormant season (winter) will not be recorded. In addition, different times of the growing season may be more important than others (i.e. May versus September) for ring width. However, in general the ring width is used to infer the overall climate change during the corresponding year (an approximation). Another problem is "memory" or autocorrelation. A stressed tree may take a year or two to recover from a hard season. This problem can be dealt with by more complex modeling (a "lag" term in the regression) or by reducing the skill estimates of chronologies.

Collection Difficulties

Tree rings must be obtained from nature, frequently from remote regions. This means that special efforts are needed to map sites properly. In addition, samples must be collected in difficult (often sloping terrain) conditions. Generally, tree rings are collected using a hand-held borer device, that requires skill to get a good sample. The best samples come from felling a tree and sectioning it. However, this requires more danger and does damage to the forest. It may not be allowed in certain areas, particularly with the oldest trees in undisturbed sites (which are the most interesting scientifically). As with all experimentalists, dendroclimatologists must, at times, decide to make the best of imperfect data, rather than resample. This tradeoff is made more difficult, because sample collection (in the field) and analysis (in the lab) may be separated significantly in time and space. These collection challenges mean that data gathering is not as simple or cheap as conventional laboratory science.

Other Measurements

Initial work focused on measuring the tree ring width—this is simple to measure and can be related to climate parameters. But the annual growth of the tree leaves other traces. In particular *maximum latewood density* (MXD) is another metric used for estimating environmental variables. It is, however, harder to measure. Other properties (e.g. isotope or chemical trace analysis) have also been tried most notably by L. M. Libby in her 1974 paper "Temperature Dependence of Isotope Ratios in Tree Rings". In theory, multiple measurements on the same ring will allow differentiation of confounding factors (e.g. precipitation and temperature). However, most studies are still based on ring widths at limiting stands.

Measuring radiocarbon concentrations in tree rings has proven to be useful in recreating past sunspot activity, with data now extending back over 11,000 years.

Synoptic Climatology

Synoptic climatology as a discipline has been continuously evolving. Unsurprisingly, even today, the field is still somewhat abstract to those first introduced to the discipline. The term 'synoptic' is by itself somewhat unspecific; in meteorology, the term is often used in reference to the synoptic scale, an area larger than the mesoscale but smaller than the continental scale, somewhere on the order of 1000–2500 km. Extending this meteorological definition would logically imply that 'synoptic climatology' is the averaging of the weather that happens on those scales; however, this definition is incomplete. Within the realm of climatology, the term 'synoptic' ('syn' ¼ same +'optic' ¼ view) is also often meant to describe viewing the atmosphere as a holistic state, a generalized overview of the important atmospheric conditions at a certain point in time Both of these perspectives on the term 'synoptic' are useful in defining the discipline of synoptic climatology however, do not fully provide an answer to what the discipline has become today. Early in the twentieth century, the fundamental building blocks of what would eventually become the discipline of synoptic climatology were discussed by Swedish climatologist Tor Bergeron. Delivering a paper in, he proposed the analysis of world climates based upon the study of larger-scale air masses, weather systems, fronts, and flow patterns – concepts that are now synonymous with the 'Bergen School' conceptual model of midlatitude weather) of which Bergeron was a part. Interestingly, these concepts were initially proposed by Bergeron as forming the field of dynamic climatology – a subfield of climatology that today is quite different. As explained, the field of synoptic climatology evolved from motives stemming from weather prediction for military needs during World War II, where Bergeron's ideas of large-scale air masses and frontal passages were key indicators of smaller-scale variables such as visibility, cloud base, and wind direction, which were important for aerial warfare over a particular region. However, prior to the computing age, making climatological sense of the meteorological data available at the time was most easily accomplished via the analysis of *categories* of weather. Categorizing the weather brings structure, order, and simplicity to an unwieldy amount of climatological data, thus allowing one to both make sense of an otherwise highly complex climate system and also more easily describe the impacts of the weather using analogues. Classification of the climate system into discrete categories also allows the climatologist to incorporate the innate range and variability of midlatitude weather into an analysis in a manner that simple climatological averages cannot.

From these roots, Barry and Perry noted various authors' historical definitions of synoptic climatology throughout the 1950s and 1960s before identifying two fundamental stages of a synoptic climatological study: a categorization of the broader-scale atmospheric circulation and an analysis of the relationship between these categories and the more local weather elements. As noted, the categorization of the atmosphere was still a key component of all synoptic climatological analyses at the time of the publication of his book in 1993; however, since 1973 (i.e., the publication of Barry and Perry's book), improved data collection and data availability in a variety of scientific disciplines had allowed for a more direct analysis of the impacts of weather on 'nonmeteorological but climatically related' surface environments. instead offered an updated definition of synoptic climatology, as a study that involves four factors: (1) A classification of atmospheric circulation, (2) A linking of two scales of analysis (the larger-scale circulation with the smaller-scale surface environment), (3) A focus on the effect of climatic *variability* on the surface environment, and (4) That the surface environment being studied is representative of a spatial region.

In the two decades that have passed since Yarnal's seminal book, the line differentiating synoptic climatology and dynamic climatology has become more defined, as the latter ventured firmly down the path of dynamic modeling of atmospheric processes and the former has expanded its focus in many ways. As we will discuss in the sections that follow, the breadth of synoptic climatological applications has expanded immensely, with extensive use in bioclimatological studies – especially those concerning human health, pollution and pollen research, and extreme events. The issue of global climate change has spurred on synoptic climatology in recent decades as well, as *synopticians* have capitalized on the vast amount of model output data produced by the modeling community, to downscale and analyze the more local-scale potential impacts of climate change.

As applications of synoptic climatology have evolved since 1993, so too has the definition of what constitutes a synoptic climatological analysis, representing an evolution in the theoretical side of the discipline as well. Today, most of the four factors that described as mandatory for a synoptic climatological analysis are largely still true but have broadened in scope. For example, one could argue that the definition of 'circulation' itself has broadened to the point of being slightly misleading, as atmospheric classifications can take many forms in addition to traditional circulation patterns – such as defining geographic spaces or multivariate entities at a single location. Further, the term 'surface environment' might more aptly be referred to simply as a 'climate-related outcome,' as these outcomes need not be at the surface nor are they always an 'environment' in as much as they are an event or occurrence. Additionally, some synoptic studies are not concerned with the actual relationship between the atmosphere and the surface in as much as using the 'surface environment' merely for statistically validating a classification.

While synoptic climatology is perhaps more unspecific today than ever, it holds onto one key defining characteristic: the classification. Contemporary synoptic climatology is a methodological perspective on climatology that creates and uses a classification of atmospheric variables (at nearly any spatial or temporal scale) to either simplify the climate system into a manageable set of states or gain a better understanding of how atmospheric variability impacts any climate-related outcome. Importantly, while synoptic climatology *as a discipline* involves the creation and application of classifications of the atmosphere, a synoptic climatological study need not *create* a classification, but merely *use* one.

With that historical perspective and working definition in mind, the sections that follow include a discussion of the major perspectives on synoptic classifications, key decisions in the research process, and a brief overview of popular applications in the recent literature. The aim of this manuscript is not necessarily to detail how a synoptic climatological research project can be undertaken (for that, please refer to or discuss specific methodologies, but rather to provide a brief overview of the theories and applications that have shaped contemporary synoptic climatology. still stands as the most complete and influential literature describing modern synoptic climatology, and thus, we draw extensively upon his work and update how the discipline has evolved since.

Key Decisions and the Human Influence in Synoptic Classifications

Researcher outlined a number of key decision points when undertaking a synoptic classification. Due to its utility in helping researchers understand the discipline, this taxonomy of classifications is still a large part of synoptic climatology today. For the sake of clarity, within this article, we will give each of these key decision points a name: *Perspective, mode, approach,* and *methodology.* Further, when we discuss a synoptic climatological study, the variables used in the classification itself will simply be called *atmospheric variables* (AVs), while the variables representing the 'surface environment' to which the classification is applied will be referenced as *outcomes.*

At the very foundation of a classification is the investigator's decision to use either a circulation-to-environment (C-to-E) or an environment-to-circulation (E-to-C) perspective. The former is a very generalizable classification of the atmosphere that is undertaken independent of the outcome for which it is to be used. One of the key benefits of using the C-to-E perspective is that the classification could be used for predictive modeling since all observations are classified. Further, the C-to-E perspective allows the existing classification to be used in additional studies – such as the Lamb weather types Grosswetterlagen type, and the spatial synoptic classification (SSC) provided that the AVs are key factors influencing the variability of the outcome within the spatial domain. For these reasons, the C-to-E perspective is much more common than its counterpart. The E-to-C perspective is more study-specific – classifying the atmosphere only when a specific outcome occurs (e.g., high ozone days, high dust days, or days when a flood occurred in.) This perspective, therefore, is limited to use only in the study at hand, since the outcome itself is predetermining which observations of the AVs are classified. Despite its unpopularity, due to the more specific focus of the E-to-C perspective on the exact outcome, location, and place of interest, it can usually offer a greater level of detail or understanding of the various atmospheric processes that lead to an outcome than the C-to-E perspective. The E-to-C perspective can be particularly effective when the outcome is an extreme event, when the binary occurrence or nonoccurrence of that event is the criterion that must be met for an observation to be classified. This perspective also defines the perspective that must be used when conducting a classic compositing analysis. Thus, each perspective provides valuable insight to the researcher.

A second key decision of synoptic classification is the mode of classifying the atmosphere. Yarnal described a number of modes, including compositing (mentioned earlier), indexing and regionalization); however, contemporary synoptic climatological classifications usually take one of two basic modes: a circulation pattern classification (CPC; sometimes referred to as a map-pattern classification) or a weather-type classification (WTC; often referred to as an air-mass classification). It is important to note that the specific terminology denoting these two modes is used loosely in various studies and has changed over time; for example, sometimes, a CPC is labeled as a WTC or its resultant patterns are called weather types. However, the difference between the two approaches is quite apparent (s-mode vs. p-mode of data decomposition; and can be discerned once reading the methodology section of any study).

The s-mode of data decomposition used for CPC is set up where each case (row in a spreadsheet) represents a different time (most commonly, a day) and each column represents a specific location (e.g., a grid point), while the spreadsheet itself represents (usually) one specific AV (e.g., 500 mb geopotential height). With this setup, each individual cell is the value of the AV at each location at each time, and the values across a whole row quantify the circulation pattern across the spatial domain for that time. Eventually, each row is classified into one of several categories based upon the similarity of the shape of that case's pattern relative to the other cases being classified. Thus, a CPC focuses on multiple locations, typically a single (occasionally multiple, e.g., AV, and its spatial variability, or pattern of flow.

CPCs have a well-developed history of classification techniques. The earliest CPCs were all very labor-intensive, requiring individual climatologists to analyze stacks of weather maps and essentially sort them into piles representing categories of either circulation patterns over a region or locations of cyclonicity. While these *manual* methods were the only option prior to the digital era, they also presented serious drawbacks such as poor reproducibility. The very first *automated* methods were also CPCs. Based upon fairly simple measures of association, these *correlation-based classifications* helped alleviate the main drawbacks with manual CPCs.

Since the mid-1990s, there has been a proliferation of climate data, the costs of increasing computing power have decreased markedly, and advanced statistical software has become relatively inexpensive and more accessible. More so than any other mode of synoptic classification, CPC has concurrently undergone a revolution, expanding tremendously in the variety of different classification methodologies and applications. Many of these methods involve principal components analysis (PCA) and cluster analysis, both of which can have multiple variants. Perhaps the newest technique to synoptic classification is that of the self-organizing map (SOM), which uses an artificial neural network for classification purposes.

Conversely to a CPC, a WTC is completed with a p-mode of data decomposition. In this mode, each case (row), again usually a day, still represents a different observation time; however, each column represents a different AV (e.g., 3 pm temperature, 3 pm dew point) and the spreadsheet typically represents one specific location. With this setup, the values across a whole row quantify a multivariate weather situation at a location on that day (somewhat akin to identifying the air mass present at a location). During the classification process, each row is classified into one of several categories based upon the similarity of that day's meteorological variables relative to those same variables on all the other days being classified. Thus, a WTC focuses most often on a single location, but multiple variables.

Often with a WTC, the classification process is repeated separately for multiple locations, thereby creating a category of surface weather conditions across an area that is representative of the synoptic-scale circulation. For example, a cold–dry anticyclonic surface weather type would likely occur on a day at a location just behind the passage of the cold front associated with northerly winds on the backside of a transient midlatitude cyclone. Much younger than CPCs, the WTC approach got its start with with the creation of the temporal synoptic index (TSI), before being modified to a spatial synoptic index, an SSC, and then finally a redeveloped SSC. In addition to using an entirely different mode of data decomposition than the CPC approach, WTCs are unique in that the *methodology* is not used nearly as often as the *resultant data*. That is, many of the studies using a WTC approach do not actually perform a classification, but rather use the categories developed from another study (most often, the redeveloped SSC;) to analyze an outcome. This said, beyond the culmination of the lineage of the WTCs described earlier, alternative WTC approaches have been undertaken, and the original WTC methodology (the TSI is still used, although not always referenced by name). Additionally, the SSC methodology has been used/modified to create that classification at new locations, while a gridded weather typing classification methodology has also recently been developed utilizing reanalysis data, thereby adding structure and spatial resolution to previous WTCs.

Another key decision point identified by is the distinction between manual and automated approaches to classification, or a variation of both (termed 'hybrid'; e.g., the SSC by). Manual methods were much more common prior to the digital age (e.g., Grosswetterlagen and the Lamb weather types). However, since Yarnal's book, most new methods of classification are automated (though not all, e.g. As part of a larger project, sorted dozens of different classification methodologies (and variants thereof) into five different categories, including four 'automated' categories: PCA-, optimization-, leader-algorithm-, and threshold-based classifications. It is important to note that the methods listed in are all generally undertaken with the CPC mode; however, this is not a requirement for automated methods (e.g., the TSI.) This recent trend towards automation is perhaps best exemplified by efforts to produce an automated version of two of the most oft-used manual classifications, the Grosswetterlagen and Lamb weather types.

Instead of the choice between manual and automated, a more recent focus in synoptic climatology has been on the evaluation of some of the human influences innate to any classification. More ubiquitous than any dilemma in the discipline is choice of the number of final categories to include in the classification. This is because the true atmosphere is dynamic in time and space, and therefore, essentially every day in the sample is unique. As part of a larger assessment, described the differences between a few different cluster number options. While there are some statistical measures that can help qualify a classification as satisfactory (in any regard, not just in terms of the number of categories) such as the pseudo F-statistic and others presented in, part of the synoptic climatologist's task is to find a balance between a manageable (i.e., not unwieldy) number of meaningful patterns and a satisfactory performance of the classification *in regard to the outcome being analyzed*. The spatially cohesive plane produced by the SOM technique has proved to be particularly helpful in this regard, as SOMs allow a larger number of categories to remain 'useful' and thus SOMs have gained popularity recently.

Beyond the number of categories, there are a handful of other choices that precede any synoptic classification. Since, numerous new sources of atmospheric data have become available, such as reanalysis data and global climate model (GCM) data, each from varying modeling centers, opening up even more choices to the investigator. With the proliferation of these data sets today and the countless options for obtaining the data, one of the first decisions is which AVs to use for classification, and the source of these AVs. Ultimately, a researcher must choose AVs that they theorize will impact the variability of the outcome. noted that sea-level pressure (SLP) is the most popular AV used historically for CPCs, though temperature and dew point generally weigh most heavily in defining WTCs. To the authors' knowledge, no *systematic* comparison of synoptic climatological classifications completed with different reanalysis products has been done, although briefly discussed differences between two products in deriving Lamb weather types and reanalysis products have been compared in other ways. There are also a multitude of GCMs available today, with the common course of action involving the evaluation of GCM ability in modeling historic circulation patterns (as represented by reanalyses) and presenting a range of future projections across multiple models and emissions scenarios. The size and resolution of the spatial domain in which the AVs are classified are also important choices. thoroughly examined a number of domain sizes in evaluating SLP patterns over Europe, while tested several different spatial resolutions of SLP in their study, noting that finer-scale resolution does not always equate to a better classification.

The treatment of the data prior to classification is also a critical factor in performing a synoptic climatological analysis. noted that many researchers using GCM data will remove any mean bias in GCM data prior to statistical downscaling – a technique that has incorporated synoptic classifications quite regularly. When using SLP data, often, the *raw* value of SLP is of less consequence than the *relative* value of SLP compared to the rest of the domain, or the gradient of change from high to low pressure centers. Thus, transformed SLP data into spatial anomalies prior to classification. If seasonality in synoptic type frequency is undesirable, it can also be reduced; using multiple moving-average filters,, for example, removed the mean seasonal cycle from 500 mb geopotential height fields prior to classification. Further, while only specific to classifications involving PCA, detailed the influence that the number of retained principal components in a PCA can have on a synoptic classification as well.

All of these human influences are in addition to selecting a final classification methodology. While a discussion on these methodologies is beyond the scope of this overview, each classification should be evaluated for effectiveness. To this end, in the mid-2000s, the European Cooperation of Science and Technology launched Action 733 (better known as COST733 in the literature) that aimed to "achieve a general numerical method for assessing, comparing and classifying weather situations in Europe". In this assessment, many of the other human influences on classification (such as the number of categories and size of domain) were also examined. One important result from the COST733 Action was the conclusion that there is no ideal classification for all purposes but that the classification needs to be tailored to the specific study at hand.

Thus, somewhat ironically, despite the 'automated' world of synoptic climatology today, these human influences are still numerous, meaning that all classifications are inherently subjective to some degree. It is these human influences that require the expertise of the synoptic climatologist.

Major Themes in Contemporary Synoptic Climatological Applications

The applications for which synoptic climatological output is used are as diverse as the methods. Dating back to the original midlatitude cyclone model developed by the Bergen School, synoptic climatologists realized the inherent connection between understanding broader-scale processes and understanding local or regional weather conditions. Within this context, a considerable amount of contemporary synoptic climatological research still aims to connect these broader-scale processes with surface weather conditions, typically temperature and precipitation.

Spatial patterns of surface air temperatures or temperature anomalies have long been correlated with synoptic patterns, such as composites of upper-level height anomalies or pressure patterns. Historically done over data-rich areas of the globe, with the addition of quality reanalysis data sets, synoptic–temperature relationships have recently begun to move into traditionally more da-ta-sparse places. Extreme temperature events are analyzed through synoptic typin, often through weather typing or compositing. Smaller-scale assessments, difficult to render in model simulations, are also evaluated using synoptics, such as the urban heat island, heat events, or cold-air drainage patterns. Other temperature metrics, such as satellite-derived land-surface temperatures (e.g., the Middle East, have also been assessed).

Precipitation patterns are commonly studied with regard to their synoptic signature as well. One of the initial applications of the types was to assess precipitation patterns related to circulation. This interest has continued with work attempting to disentangle the contributions to total precipitation from different synoptic types, often with the aim of evaluating causal mechanisms or trends. As with temperature, this has evolved over time to encompass an assessment of climate variability across more remote locations and more data sources, such as radar-estimated rainfall and satellite-estimated rainfall. Research has focused on systems on various scales, from the West African monsoon to sea breeze-generated convection in Java. Given their significant impact, extreme and unusual events have been frequently assessed using the synoptic methodology, such as floods, heavy snowfall, drought, and freezing rain. Addressing aspects of storminess and circulation, synoptic methods have associated circulation with tornado incidence, extreme waves, and water clarity.

Another common theme of recent synoptic research is its use with weather and climate model output. Starting with many papers have subsequently validated GCM ability to represent the full range of atmospheric flow patterns, and multiple models have been compared in terms of their relative abilities to replicate patterns both individually and as an ensemble. These methods are also used to examine very specific phenomena that are difficult to model, such as, which evaluates US Great Lake effect snow simulation, or, which evaluates reanalysis ability to simulate flows in the Ross Sea region of Antarctica. Synoptic methods can also be employed as a method of statistical downscaling with model output, including analog methods in which historical patterns are matched with patterns in a model, or through regression or neural network models on principal components. Further, some studies using synoptic climatology to downscale precipitation find that using synoptic patterns can result in more realistic precipitation fields than using GCM-generated precipitation data.

Since the 1990s, considerable research has examined variability in teleconnections and synoptic climatological patterns. Some works, such as, use synoptic climatology to demonstrate the ability of rendering certain phenomena within a synoptic framework (in their case, SOMs), and as noted earlier, this often is part of model validation studies as well. More commonly, the relationship between teleconnection indices and climate system response in a specific region is explored, where frequencies of different patterns are associated with teleconnection phases; this is then often used to assess relationships with climatic variables such as precipitation or water resources.

Given the large body of research devoted to climate change, many papers utilize synoptic methods to assess historical pattern changes as indicators of climate change; this has been especially true in exploring the relative contribution of differences in circulation pattern frequencies compared with 'within-pattern' warming.with most research suggesting that pattern frequency changes are not the prime driver of climate trends.

Air quality has strong connections with atmospheric conditions, as the holistic suite of atmospheric conditions – inclusive of thermal conditions, atmospheric stability, radiation, wind, and precipitation – play a role. The holistic nature of synoptic climatology allows these conditions to be evaluated in concert, and so it is especially useful for air pollution or pollen transport. Both historical and more recent studies have used the synoptic method to assess differences in pollution levels across circulation patterns, between rural and urban environments, as well as to determine source regions for pollutants. Pollen dispersion can also be correlated well with synoptic conditions.

More broadly, it is clear that all life interacts with, and is affected by, the holistic set of weather conditions; it is hence unsurprising that synoptic climatology has been used in a variety of biometeorological assessments as well. A number of methods, most commonly WTC, stem from the aforementioned connection with air quality itself, such as who used air quality–synoptic relationships and then related these to hospital admissions. Another common area is extreme temperature-related mortality, in which weather types and circulation patterns measure variability by synoptic type; other studies such asextend weather conditions more broadly to hospitaliza- tions. While human health-based research using WTC methods is quite prevalent, nonhuman biometeo- rological studies are also undertaken, including those analyzing wine and insect transport.

Synoptic climatology has evolved markedly since the discipline was first conceived in the 1940s. The resurgence of the applied side of the field over the last two decades is particularly encouraging. At a time when climatology has become entrenched in the international scientific, political, and public forums, synoptic climatology has helped shape our understanding of how the atmosphere interacts with the Earth system as whole. However, despite the wide-ranging *applications* discussed herein, synoptic climatology is still underutilized in many arenas (e.g., oceans, ecosystems, and agriculture) and, thus, still has potential for more widespread utility. There is also room for growth in the realm of classification *methodologies*. In particular, even the more recently developed classifications have largely underutilized the amount of climate data and computing power available today. To this end, new classification modes that identify synoptic types in greater than two dimensions (e.g., three- or four-dimensional types using latitude, longitude, time, multiple vertical levels, and multiple AVs) may be a natural progression.

Hydroclimatology

Hydroclimatology provides a systematic structure for analyzing how the climate system causes time and space variations (both global and local) in the hydrologic cycle. Changes in the relationship between the climate system and the hydrologic cycle underlie floods, drought and possible future influences of global warming on water resources.

Hydroclimatology was defined by Walter Langbein as the 'study of the influence of climate upon the waters of the land'. It includes hydrometeorology as well as the surface and near-surface water processes of evaporation, runoff, groundwater recharge and interception.

Hydroclimatology is the study of moisture in the atmosphere and water in and on the surface of the earth in all its various forms. Thus, hydroclimatology is that area of research where the fields of climatology and hydrology intersect. It includes fluxes of moisture between the atmosphere and the surface (i.e., precipitation and evapotranspiration), atmospheric water vapor and its circulation both horizontally and vertically, water stored on the land (snow cover, glaciers, lakes, and streams) and on the oceans (sea ice), water stored in the soil (soil moisture), and water used by vegetation.

Climate science applied to freshwater reservoirs – surface hydrology, groundwater, and the cryosphere – and the resulting impact on society is a primary component of the growing field of hydroclimatology.

Hydroclimatology is the interaction between climate and hydrology, e.g precipitation, evapotranspiration, groundwater, rivers, lakes, water balance, energy balance and describes the energy and moisture exchanges between atmosphere, surface, and subsurface.

Other Sub-divisions

Applied Climatology

As with applied chemistry, applied physics and so on, applied climatology is about studying what is actually happening now rather than climate theory of what will happen, or the use of theoretical models to predict events in the short or long-term future. This means that applied climatology has much in common with some of the other atmospheric sciences such as meteorology. Climate events that are happening now have immediate and measurable impacts on weather systems - locally and far away. It is arguably the oldest form of climatology, originating long before the birth of the modern science. Military decisions were made based on climate events and weather patterns, and Ancient Egypt's system of agriculture was based on the observation of the fluctuations in the Nile's "inundation" (flood waters bringing nutrient-rich silt that was the backbone of their agricultural surplus) allowed for long-term planning of social, agricultural and civic project planning.

Boundary-Layer Climatology

The Boundary Layer is the lowest level of the atmosphere, the area most affected by local and planetary climate change. This is the atmospheric layer that experiences the most turbulence and holds the important weather systems, interacting with the "thermals", distributing air and moisture across the planet. Therefore, boundary layer climatologists study the networks and interactions of the lower atmosphere, and their impact on weather and climate systems, but also how the network of winds are affected by such climate phenomena as rising air and sea temperatures, urban heat islands, and natural events such as volcanic activity. It also has critical uses for immediate weather patterns in meteorology. Neither boundary-layer climatology nor meteorology exist in isolation and are intrinsically linked.

Dynamic Climatology

No science exists in a bubble and no subdivision of a broad science does so either. Dynamic climatologists are concerned with examining the accumulated sum total of information acquired from all related sciences, typically quantitative in nature, based on observed phenomena. Dynamic climatology examines and handles everything from paleodata to volcanic eruption to looking at short range or short-term weather patterns to long-term climate effects from natural or anthropogenic causes. This area of climatology uses a holistic approach.

Physical Climatology

Most of climatology is concerned with looking at data and making projections or presenting the data as facts, statistics, graphs and hard figures - it is quantitative in nature. Physical climatology is more qualitative. It examines and explains how climate can shape topography and geographical systems. For example, it seeks to explain how glaciation is one factor capable of forming valleys and mountains, how extreme flooding events will change a landscape. For example, we know that the Vlad - of North America's largest glaciers created the enormous basin around the Great Lakes and then filled it with meltwater to create the lakes we have today.

References

- What-paleoclimatology, news: ncdc.noaa.gov, Retrieved 16 January, 2019

- "are you ready?". Federal emergency management agency. 2006-04-05. Archived from the original on 2006-06-29. Retrieved 2006-06-24

- "dennis, katrina, rita, stan, and wilma "retired" from list of storm names." noaa. Retrieved on june 14, 2008

- Climatology: environmentalscience.org, Retrieved 30 April, 2019

- E.b. Rodgers and r.f. Adler. Tropical cyclone rainfall characteristics as determined from a satellite passive microwave radiometer. Retrieved on 2008-04-16

- Bioclimatology, science: britannica.com, Retrieved 13 August, 2019

- Mills, g (2007). "luke howard and the climate of london". Weather. 63 (6): 153–157. Bibcode:2008wthr...63..153m. Doi:10.1002/wea.195

- Hydroclimatology-definition: hydroclimatology.info, Retrieved 15 July, 2019

- Surface temperature reconstructions for the last 2,000 years. Washington, d.c: national academies press. 2006. Isbn 0-309-10225-1

Climate Change

4

- **Climate Sensitivity**

- **Global Warming**

- **Greenhouse Effect**

- **Climate Change Adaptation**

- **Effects of Climate Change**

- **Climate Change Mitigation**

- **Climate Engineering**

Climate change refers to the changes in the climate system of the Earth that result in new weather patterns that remain the same for a long period of time. Some of the major areas of study within this field are global warming and greenhouse effect. The topics elaborated in this chapter will help in developing a thorough understanding of climate change, its effects and the ways to mitigate them.

Climate change is the periodic modification of Earth's climate brought about as a result of changes in the atmosphere as well as interactions between the atmosphere and various other geologic, chemical, biological, and geographic factors within the Earth system.

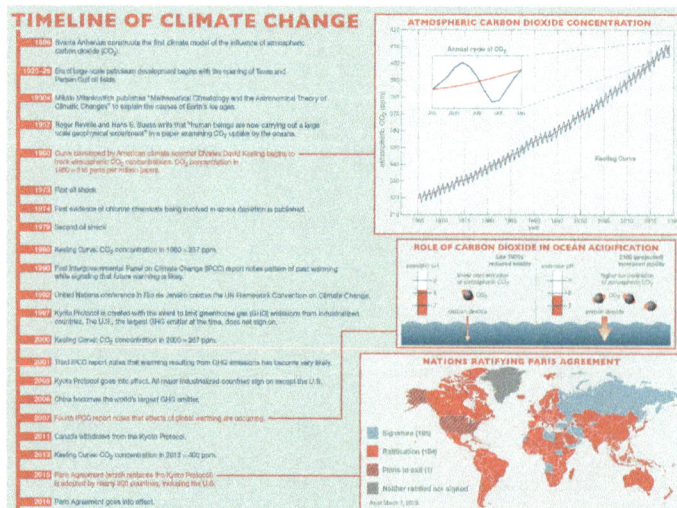

A timeline of important developments in climate change.

The atmosphere is a dynamic fluid that is continually in motion. Both its physical properties and its rate and direction of motion are influenced by a variety of factors, including solar radiation, the geographic position of continents, ocean currents, the location and orientation of mountain ranges, atmospheric chemistry, and vegetation growing on the land surface. All these factors change through time. Some factors, such as the distribution of heat within the oceans, atmospheric chemistry, and surface vegetation, change at very short timescales. Others, such as the position of continents and the location and height of mountain ranges, change over very long timescales. Therefore, climate, which results from the physical properties and motion of the atmosphere, varies at every conceivable timescale.

Climate is often defined loosely as the average weather at a particular place, incorporating such features as temperature, precipitation, humidity, and windiness. A more specific definition would state that climate is the mean state and variability of these features over some extended time period. Both definitions acknowledge that the weather is always changing, owing to instabilities in the atmosphere. And as weather varies from day to day, so too does climate vary, from daily day-and-night cycles up to periods of geologic time hundreds of millions of years long. In a very real sense, climate variation is a redundant expression—climate is always varying. No two years are exactly alike, nor are any two decades, any two centuries, or any two millennia.

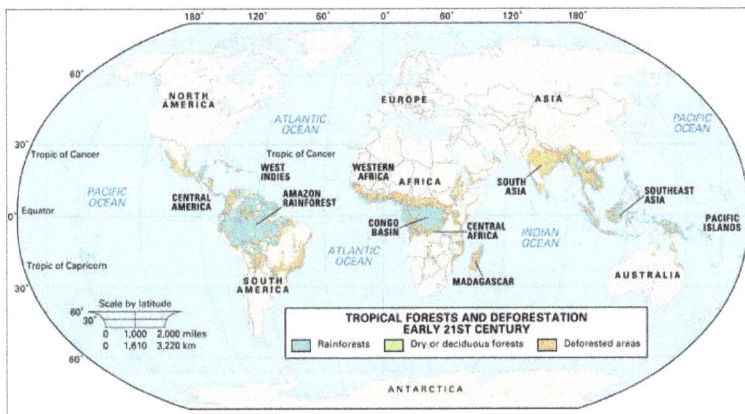

Tropical forests and deforestation Tropical forests and deforestation in the early 21st century.

Climate Change within a Human Life Span

Regardless of their locations on the planet, all humans experience climate variability and change within their lifetimes. The most familiar and predictable phenomena are the seasonal cycles, to which people adjust their clothing, outdoor activities, thermostats, and agricultural practices. However, no two summers or winters are exactly alike in the same place; some are warmer, wetter, or stormier than others. This interannual variation in climate is partly responsible for year-to-year variations in fuel prices, crop yields, road maintenance budgets, and wildfire hazards. Single-year, precipitation-driven floods can cause severe economic damage, such as those of the upper Mississippi River drainage basin during the summer of 1993, and loss of life, such as those that devastated much of Bangladesh in the summer of 1998. Similar damage and loss of life can also occur as the result of wildfires, severe storms, hurricanes, heat waves, and other climate-related events.

Climate variation and change may also occur over longer periods, such as decades. Some locations experience multiple years of drought, floods, or other harsh conditions. Such decadal variation of climate poses challenges to human activities and planning. For example, multiyear droughts can disrupt water supplies, induce crop failures, and cause economic and social dislocation, as in the case of the Dust Bowl droughts in the midcontinent of North America during the 1930s. Multiyear droughts may even cause widespread starvation, as in the Saheldrought that occurred in northern Africa during the 1970s and '80s.

Seasonal Variation

Every place on Earth experiences seasonal variation in climate (though the shift can be slight in some tropical regions). This cyclic variation is driven by seasonal changes in the supply of solar radiation to Earth's atmosphere and surface. Earth's orbit around the Sun is elliptical; it is closer to the Sun (147 million km about 91 million miles) near the winter solstice and farther from the Sun (152 million km about 94 million miles) near the summer solstice in the Northern Hemisphere. Furthermore, Earth's axis of rotation occurs at an oblique angle (23.5°) with respect to its orbit. Thus, each hemisphere is tilted away from the Sun during its winter period and toward the Sun in its summer period. When a hemisphere is tilted away from the Sun, it receives less solar radiation than the opposite hemisphere, which at that time is pointed toward the Sun. Thus, despite the closer proximity of the Sun at the winter solstice, the Northern Hemisphere receives less solar radiation during the winter than it does during the summer. Also as a consequence of the tilt, when the Northern Hemisphere experiences winter, the Southern Hemisphere experiences summer.

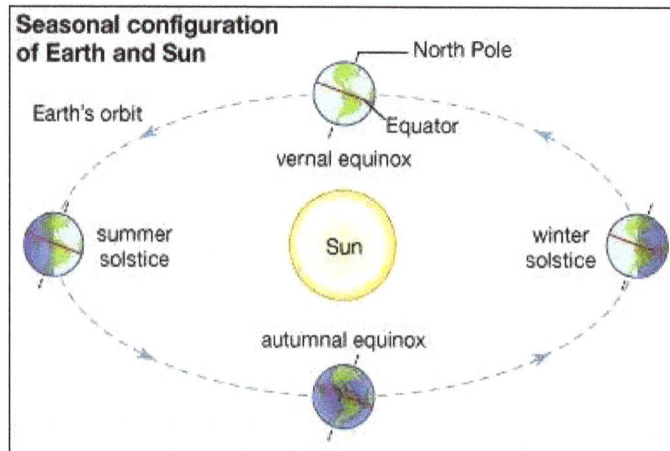

A diagram shows the position of Earth at the beginning of each season in the Northern Hemisphere.

Earth's climate system is driven by solar radiation; seasonal differences in climate ultimately result from the seasonal changes in Earth's orbit. The circulation of air in the atmosphere and water in the oceans responds to seasonal variations of available energy from the Sun. Specific seasonal changes in climate occurring at any given location on Earth's surface largely result from the transfer of energy from atmospheric and oceanic circulation. Differences in surface heating taking place between summer and winter cause storm tracks and pressure centres to shift position and strength. These heating differences also drive seasonal changes in cloudiness, precipitation, and wind.

Seasonal responses of the biosphere (especially vegetation) and cryosphere (glaciers, sea ice, snowfields) also feed into atmospheric circulation and climate. Leaf fall by deciduous trees as they go into winter dormancy increases the albedo (reflectivity) of Earth's surface and may lead to greater local and regional cooling. Similarly, snow accumulation also increases the albedo of land surfaces and often amplifies winter's effects.

Interannual Variation

Interannual climate variations, including droughts, floods, and other events, are caused by a complex array of factors and Earth system interactions. One important feature that plays a role in these variations is the periodic change of atmospheric and oceanic circulation patterns in the tropical Pacific region, collectively known as El Niño–Southern Oscillation (ENSO) variation. Although its primary climatic effects are concentrated in the tropical Pacific, ENSO has cascading effects that often extend to the Atlantic Ocean region, the interior of Europe and Asia, and the polar regions. These effects, called teleconnections, occur because alterations in low-latitude atmospheric circulation patterns in the Pacific region influence atmospheric circulation in adjacent and downstream systems. As a result, storm tracks are diverted and atmospheric pressure ridges (areas of high pressure) and troughs (areas of low pressure) are displaced from their usual patterns.

As an example, El Niño events occur when the easterly trade winds in the tropical Pacific weaken or reverse direction. This shuts down the upwelling of deep, cold waters off the west coast of South America, warms the eastern Pacific, and reverses the atmospheric pressure gradient in the western Pacific. As a result, air at the surface moves eastward from Australiaand Indonesia toward the central Pacific and the Americas. These changes produce high rainfall and flash floods along the normally arid coast of Peru and severe drought in the normally wet regions of northern Australia and Indonesia. Particularly severe El Niño events lead to monsoon failure in the Indian Ocean region, resulting in intense drought in India and East Africa. At the same time, the westerlies and storm tracks are displaced toward the Equator, providing California and the desert Southwest of the United States with wet, stormy winter weather and causing winter conditions in the Pacific Northwest, which are typically wet, to become warmer and drier. Displacement of the westerlies also results in drought in northern China and from northeastern Brazil through sections of Venezuela. Long-term records of ENSO variation from historical documents, tree rings, and reef corals indicate that El Niño events occur, on average, every two to seven years. However, the frequency and intensity of these events vary through time.

The North Atlantic Oscillation (NAO) is another example of an interannual oscillation that produces important climatic effects within the Earth system and can influence climate throughout the Northern Hemisphere. This phenomenon results from variation in the pressure gradient, or the difference in atmospheric pressure between the subtropical high, usually situated between the Azores and Gibraltar, and the Icelandic low, centred between Iceland and Greenland. When the pressure gradient is steep due to a strong subtropical high and a deep Icelandic low (positive phase), northern Europe and northern Asia experience warm, wet winters with frequent strong winter storms. At the same time, southern Europe is dry. The eastern United States also experiences warmer, less snowy winters during positive NAO phases, although the effect is not as great as in Europe. The pressure gradient is dampened when NAO is in a negative mode—that is, when a weaker pressure gradient exists from the presence of a weak subtropical high and Icelandic low. When this happens, the Mediterranean region receives abundant winter rainfall, while northern Europe is cold and dry. The eastern United States is typically colder and snowier during a negative NAO phase.

During years when the North Atlantic Oscillation (NAO) is in its positive phase, the eastern United States, southeastern Canada, and northwestern Europe experience warmer winter temperatures, whereas colder temperatures are found in these locations during its negative phase. When the El Niño/Southern Oscillation (ENSO) and NAO are both in their positive phase, European winters tend to be wetter and less severe; however, beyond this general tendency, the influence of the ENSO upon the NAO is not well understood.

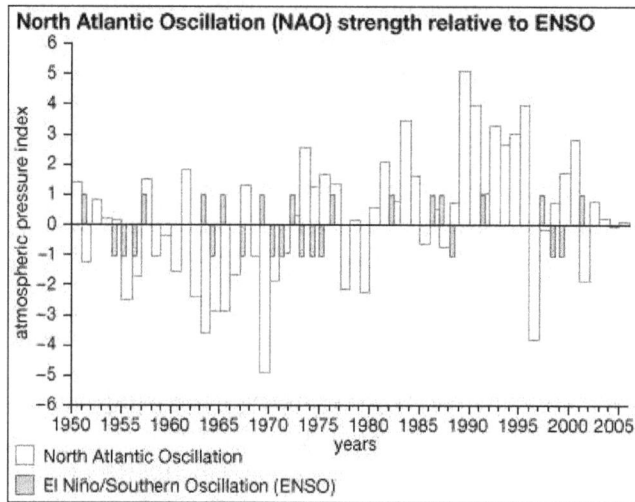

The ENSO and NAO cycles are driven by feedbacks and interactions between the oceans and atmosphere. Interannual climate variation is driven by these and other cycles, interactions among cycles, and perturbations in the Earth system, such as those resulting from large injections of aerosols from volcanic eruptions. One example of a perturbation due to volcanismis the 1991 eruption of Mount Pinatubo in the Philippines, which led to a decrease in the average global temperature of approximately 0.5 °C (0.9 °F) the following summer.

Decadal Variation

Climate varies on decadal timescales, with multiyear clusters of wet, dry, cool, or warm conditions. These multiyear clusters can have dramatic effects on human activities and welfare. For instance, a severe three-year drought in the late 16th century probably contributed to the destruction of Sir Walter Raleigh's "Lost Colony" at Roanoke Island in what is now North Carolina, and a subsequent seven-year drought led to high mortality at the Jamestown Colony in Virginia. Also, some scholars have implicated persistent and severe droughts as the main reason for the collapse of the Maya civilization in Mesoamerica between AD 750 and 950; however, discoveries in the early 21st century suggest that war-related trade disruptions played a role, possibly interacting with famines and other drought-related stresses.

Although decadal-scale climate variation is well documented, the causes are not entirely clear. Much decadal variation in climate is related to interannual variations. For example, the frequency and magnitude of ENSO change through time. The early 1990s were characterized by repeated El Niño events, and several such clusters have been identified as having taken place during the 20th century. The steepness of the NAO gradient also changes at decadal timescales; it has been particularly steep since the 1970s.

Recent research has revealed that decadal-scale variations in climate result from interactions between the ocean and the atmosphere. One such variation is the Pacific Decadal Oscillation (PDO), also referred to as the Pacific Decadal Variability (PDV), which involves changing sea surface temperatures (SSTs) in the North Pacific Ocean. The SSTs influence the strength and position of the Aleutian Low, which in turn strongly affects precipitation patterns along the Pacific Coast of North America. PDO variation consists of an alternation between "cool-phase" periods, when coastal Alaska is relatively dry and the Pacific Northwest relatively wet, and "warm-phase" periods, characterized by relatively high precipitation in coastal Alaska and low precipitation in the Pacific Northwest. Tree ring and coral records, which span at least the last four centuries, document PDO variation.

A similar oscillation, the Atlantic Multidecadal Oscillation (AMO), occurs in the North Atlantic and strongly influences precipitation patterns in eastern and central North America. A warm-phase AMO (relatively warm North Atlantic SSTs) is associated with relatively high rainfall in Florida and low rainfall in much of the Ohio Valley. However, the AMO interacts with the PDO, and both interact with interannual variations, such as ENSO and NAO, in complex ways. Such interactions may lead to the amplification of droughts, floods, or other climatic anomalies. For example, severe droughts over much of the conterminous United States in the first few years of the 21st century were associated with warm-phase AMO combined with cool-phase PDO. The mechanisms underlying decadal variations, such as PDO and AMO, are poorly understood, but they are probably related to ocean-atmosphere interactions with larger time constants than interannual variations. Decadal climatic variations are the subject of intense study by climatologists and paleoclimatologists.

Climate Change since the Emergence of Civilization

Human societies have experienced climate change since the development of agriculture some 10,000 years ago. These climate changes have often had profound effects on human culturesand societies. They include annual and decadal climate fluctuations such as those described, as well as large-magnitude changes that occur over centennial to multimillennial timescales. Such changes are believed to have influenced and even stimulated the initial cultivation and domestication of crop plants, as well as the domestication and pastoralization of animals. Human societies have changed adaptively in response to climate variations, although evidence abounds that certain societies and civilizations have collapsed in the face of rapid and severe climatic changes.

Centennial-scale Variation

Historical records as well as proxy records (particularly tree rings, corals, and ice cores) indicate that climate has changed during the past 1,000 years at centennial timescales; that is, no two centuries have been exactly alike. During the past 150 years, the Earth system has emerged from a period called the Little Ice Age, which was characterized in the North Atlantic regionand elsewhere by relatively cool temperatures. The 20th century in particular saw a substantial pattern of warming in many regions. Some of this warming may be attributable to the transition from the Little Ice Age or other natural causes. However, many climate scientists believe that much of the 20th-century warming, especially in the later decades, resulted from atmospheric accumulation of greenhouse gases (especially carbon dioxide, CO_2).

The Little Ice Age is best known in Europe and the North Atlantic region, which experienced relatively cool conditions between the early 14th and mid-19th centuries. This was not a period of uniformly cool climate, since interannual and decadal variability brought many warm years. Furthermore, the coldest periods did not always coincide among regions; some regions experienced relatively warm conditions at the same time others were subjected to severely cold conditions. Alpine glaciers advanced far below their previous (and present) limits, obliterating farms, churches, and villages in Switzerland, France, and elsewhere. Frequent cold winters and cool, wet summers ruined wine harvests and led to crop failures and famines over much of northern and central Europe. The North Atlantic cod fisheries declined as ocean temperatures fell in the 17th century. The Norse colonies on the coast of Greenland were cut off from the rest of Norse civilization during the early 15th century as pack ice and storminess increased in the North Atlantic. The western colony of Greenland collapsed through starvation, and the eastern colony was abandoned. In addition, Iceland became increasingly isolated from Scandinavia.

The Little Ice Age was preceded by a period of relatively mild conditions in northern and central Europe. This interval, known as the Medieval Warm Period, occurred from approximately AD 1000 to the first half of the 13th century. Mild summers and winters led to good harvests in much of Europe. Wheat cultivation and vineyards flourished at far higher latitudes and elevations than today. Norse colonies in Iceland and Greenland prospered, and Norse parties fished, hunted, and explored the coast of Labrador and Newfoundland. The Medieval Warm Period is well documented in much of the North Atlantic region, including ice cores from Greenland. Like the Little Ice Age, this time was neither a climatically uniform period nor a period of uniformly warm temperatures everywhere in the world. Other regions of the globe lack evidence for high temperatures during this period.

Much scientific attention continues to be devoted to a series of severe droughts that occurred between the 11th and 14th centuries. These droughts, each spanning several decades, are well documented in tree-ring records across western North America and in the peatland records of the Great Lakes region. The records appear to be related to ocean temperature anomalies in the Pacific and Atlantic basins, but they are still inadequately understood. The information suggests that much of the United States is susceptible to persistent droughts that would be devastating for water resources and agriculture.

Millennial and Multimillennial Variation

The climatic changes of the past thousand years are superimposed upon variations and trends at both millennial timescales and greater. Numerous indicators from eastern North America and Europe show trends of increased cooling and increased effective moisture during the past 3,000 years. For example, in the Great Lakes–St. Lawrence regions along the U.S.-Canadian border, water levels of the lakes rose, peatlands developed and expanded, moisture-loving trees such as beech and hemlock expanded their ranges westward, and populations of boreal trees, such as spruce and tamarack, increased and expanded southward. These patterns all indicate a trend of increased effective moisture, which may indicate increased precipitation, decreased evapotranspiration due to cooling, or both. The patterns do not necessarily indicate a monolithic cooling event; more complex climatic changes probably occurred. For example, beech expanded northward and spruce southward during the past 3,000 years in both eastern North America and western Europe. The beech expansions may indicate milder winters or longer growing seasons, whereas the spruce expansions appear related to cooler, moister summers. Paleoclimatologists are applying a variety of approaches and proxies to help identify such changes in seasonal temperature and moisture during the Holocene Epoch.

Just as the Little Ice Age was not associated with cool conditions everywhere, so the cooling and moistening trend of the past 3,000 years was not universal. Some regions became warmer and drier during the same time period. For example, northern Mexico and the Yucatanexperienced decreasing moisture in the past 3,000 years. Heterogeneity of this type is characteristic of climatic change, which involves changing patterns of atmospheric circulation. As circulation patterns change, the transport of heat and moisture in the atmosphere also changes. This fact explains the apparent paradox of opposing temperature and moisture trends in different regions.

The trends of the past 3,000 years are just the latest in a series of climatic changes that occurred over the past 11,700 years or so—the interglacial period referred to as the Holocene Epoch. At the start of the Holocene, remnants of continental glaciers from the last glaciationstill covered much of eastern and central Canada and parts of Scandinavia. These ice sheets largely disappeared by 6,000 years ago. Their absence— along with increasing sea surface temperatures, rising sea levels (as glacial meltwater flowed into the world's oceans), and especially changes in the radiation budget of Earth's surface owing to Milankovitch variations (changes in the seasons resulting from periodic adjustments of Earth's orbit around the Sun)—affected atmospheric circulation. The diverse changes of the past 10,000 years across the globe are difficult to summarize in capsule, but some general highlights and large-scale patterns are worthy of note. These include the presence of early to mid-Holocene thermal maxima in various locations, variation in ENSO patterns, and an early to mid-Holocene amplification of the Indian Ocean monsoon.

Thermal Maxima

Many parts of the globe experienced higher temperatures than today some time during the early to mid-Holocene. In some cases the increased temperatures were accompanied by decreased moisture availability. Although the thermal maximum has been referred to in North America and elsewhere as a single widespread event (variously referred to as the "Altithermal," "Xerothermic Interval," "Climatic Optimum," or "Thermal Optimum"), it is now recognized that the periods of maximum temperatures varied among regions. For example, northwestern Canada experienced its highest temperatures several thousand years earlier than central or eastern North America. Similar heterogeneity is seen in moisture records. For instance, the record of the prairie-forest boundary in the Midwestern region of the United States shows eastward expansion of prairie in Iowa and Illinois 6,000 years ago (indicating increasingly dry conditions), whereas Minnesota's forests expanded westward into prairie regions at the same time (indicating increasing moisture). The Atacama Desert, located primarily in present-day Chile and Bolivia, on the western side of South America, is one of the driest places on Earth today, but it was much wetter during the early Holocene when many other regions were at their driest.

The primary driver of changes in temperature and moisture during the Holocene was orbital variation, which slowly changed the latitudinal and seasonal distribution of solar radiation on Earth's surface and atmosphere. However, the heterogeneity of these changes was caused by changing patterns of atmospheric circulation and ocean currents.

ENSO Variation in the Holocene

Because of the global importance of ENSO variation today, Holocene variation in ENSO patterns and intensity is under serious study by paleoclimatologists. The record is still fragmentary, but evidence from fossil corals, tree rings, lake records, climate modeling, and other approaches is accumulating that suggests that (1) ENSO variation was relatively weak in the early Holocene, (2) ENSO has undergone centennial to millennial variations in strength during the past 11,700 years, and (3) ENSO patterns and strength similar to those currently in place developed within the past 5,000 years. This evidence is particularly clear when comparing ENSO variation over the past 3,000 years to today's patterns. The causes of long-term ENSO variation are still being explored, but changes in solar radiation owing to Milankovitch variations are strongly implicated by modeling studies.

Amplification of the Indian Ocean Monsoon

Much of Africa, the Middle East, and the Indian subcontinent are under the strong influence of an annual climatic cycle known as the Indian Ocean monsoon. The climate of this region is highly seasonal, alternating between clear skies with dry air (winter) and cloudy skies with abundant rainfall (summer). Monsoon intensity, like other aspects of climate, is subject to interannual, decadal, and centennial variations, at least some of which are related to ENSO and other cycles. Abundant evidence exists for large variations in monsoon intensity during the Holocene Epoch. Paleontological and paleoecological studies show that large portions of the region experienced much greater precipitation during the early Holocene than today. Lake and wetland sediments dating to this period have been found under the sands of parts of the Sahara Desert. These sediments contain fossils of elephants, crocodiles, hippopotamuses, and giraffes, together with pollen evidence of forest and woodland vegetation. In arid and semiarid parts of Africa, Arabia, and India, large and deep freshwater lakes occurred in basins that are now dry or are occupied by shallow, saline lakes. Civilizations based on plant cultivation and grazing animals, such as the Harappan civilization of northwestern India and adjacent Pakistan, flourished in these regions, which have since become arid.

These and similar lines of evidence, together with paleontological and geochemical data from marine sediments and climate-modeling studies, indicate that the Indian Ocean monsoon was greatly amplified during the early Holocene, supplying abundant moisture far inland into the African and Asian continents. This amplification was driven by high solar radiation in summer, which was approximately 7 percent higher 11,700 years ago than today and resulted from orbital forcing (changes in Earth's eccentricity, precession, and axial tilt). High summer insolation resulted in warmer summer air temperatures and lower surface pressure over continental regions and, hence, increased inflow of moisture-laden air from the Indian Ocean to the continental interiors. Modeling studies indicate that the monsoonal flow was further amplified by feedbacks involving the atmosphere, vegetation, and soils. Increased moisture led to wetter soils and lusher vegetation, which in turn led to increased precipitation and greater penetration of moist air into continental interiors. Decreasing summer insolation during the past 4,000–6,000 years led to the weakening of the Indian Ocean monsoon.

Climate Change Since the Advent of Humans

The history of humanity—from the initial appearance of genus *Homo* over 2,000,000 years ago to the advent and expansion of the modern human species (*Homo sapiens*) beginning some 150,000 years ago—is integrally linked to climate variation and change. *Homo sapiens* has experienced nearly two full glacial-interglacial cycles, but its global geographical expansion, massive population increase, cultural diversification, and worldwide ecological domination began only during the last glacial period and accelerated during the last glacial-interglacial transition. The first bipedal apes appeared in a time of climatic transition and variation, and *Homo erectus*, an extinct species possibly ancestral to modern humans, originated during the colder Pleistocene Epoch and survived both the transition period and multiple glacial-interglacial cycles. Thus, it can be said that climate variation has been the midwife of humanity and its various cultures and civilizations.

Recent Glacial and Interglacial Periods

The Most Recent Glacial Phase

With glacial ice restricted to high latitudes and altitudes, Earth 125,000 years ago was in an interglacial period similar to the one occurring today. During the past 125,000 years, however, the Earth system went through an entire glacial-interglacial cycle, only the most recent of many taking place over the last million years. The most recent period of cooling and glaciationbegan approximately 120,000 years ago. Significant ice sheets developed and persisted over much of Canada and northern Eurasia.

After the initial development of glacial conditions, the Earth system alternated between two modes, one of cold temperatures and growing glaciers and the other of relatively warm temperatures (although much cooler than today) and retreating glaciers. These Dansgaard-Oeschger (DO) cycles, recorded in both ice cores and marine sediments, occurred approximately every 1,500 years. A lower-frequency cycle, called the Bond cycle, is superimposed on the pattern of DO cycles; Bond cycles occurred every 3,000–8,000 years. Each Bond cycle is characterized by unusually cold conditions that take place during the cold phase of a DO cycle, the subsequent Heinrich event (which is a brief dry and cold phase), and the rapid warming phase that follows each Heinrich event. During each Heinrich event, massive fleets of icebergs were released into the North Atlantic, carrying rocks picked up by the glaciers far out to sea. Heinrich events are marked in marine sediments by conspicuous layers of iceberg-transported rock fragments.

Many of the transitions in the DO and Bond cycles were rapid and abrupt, and they are being studied intensely by paleoclimatologists and Earth system scientists to understand the driving mechanisms of such dramatic climatic variations. These cycles now appear to result from interactions between the atmosphere, oceans, ice sheets, and continental rivers that influence thermohaline circulation (the pattern of ocean currents driven by differences in water density, salinity, and temperature, rather than wind). Thermohaline circulation, in turn, controls ocean heat transport, such as the Gulf Stream.

The Last Glacial Maximum

During the past 25,000 years, the Earth system has undergone a series of dramatic transitions. The most recent glacial period peaked 21,500 years ago during the Last Glacial Maximum, or LGM. At that time, the northern third of North America was covered by the Laurentide Ice Sheet, which extended as far south as Des Moines, Iowa; Cincinnati, Ohio; and New York City. The Cordilleran Ice Sheet covered much of western Canada as well as northern Washington, Idaho, and Montana in the United States. In Europe the Scandinavian Ice Sheet sat atop the British Isles, Scandinavia, northeastern Europe, and north-central Siberia. Montane glaciers were extensive in other regions, even at low latitudes in Africa and South America. Global sea level was 125 metres (410 feet) below modern levels, because of the long-term net transfer of water from the oceans to the ice sheets. Temperatures near Earth's surface in unglaciated regions were about 5 °C (9 °F) cooler than today. Many Northern Hemisphere plant and animal species inhabited areas far south of their present ranges. For example, jack pine and white spruce trees grew in northwestern Georgia, 1,000 km (600 miles) south of their modern range limits in the Great Lakes region of North America.

The Last Deglaciation

The continental ice sheets began to melt back about 20,000 years ago. Drilling and dating of submerged fossil coral reefs provide a clear record of increasing sea levels as the ice melted. The most rapid melting began 15,000 years ago. For example, the southern boundary of the Laurentide Ice Sheet in North America was north of the Great Lakes and St. Lawrence regions by 10,000 years ago, and it had completely disappeared by 6,000 years ago.

The warming trend was punctuated by transient cooling events, most notably the Younger Dryas climate interval of 12,800–11,600 years ago. The climatic regimes that developed during the deglaciation period in many areas, including much of North America, have no modern analog (i.e., no regions exist with comparable seasonal regimes of temperature and moisture). For example, in the interior of North America, climates were much more continental (that is, characterized by warm summers and cold winters) than they are today. Also, paleontological studies indicate assemblages of plant, insect, and vertebrate species that do not occur anywhere today. Spruce trees grew with temperate hardwoods (ash, hornbeam, oak, and elm) in the upper Mississippi River and Ohio River regions. In Alaska, birch and poplar grew in woodlands, and there were very few of the spruce trees that dominate the present-day Alaskan landscape. Boreal and temperate mammals, whose geographic ranges are widely separated today, coexisted in central North America and Russia during this period of deglaciation. These unparalleled climatic conditions probably resulted from the combination of a unique orbital pattern that increased summer insolation and reduced winter insolation in the Northern Hemisphere and the continued presence of Northern Hemisphere ice sheets, which themselves altered atmospheric circulation patterns.

Climate Change and the Emergence of Agriculture

The first known examples of animal domestication occurred in western Asia between 11,000 and 9,500 years ago when goats and sheep were first herded, whereas examples of plant domestication date to 9,000 years ago when wheat, lentils, rye, and barley were first cultivated. This phase of technological increase occurred during a time of climatic transition that followed the last glacial period. A number of scientists have suggested that, although climate change imposed stresses on hunter-gatherer-forager societies by causing rapid shifts in resources, it also provided opportunities as new plant and animal resources appeared.

Glacial and Interglacial Cycles of the Pleistocene

The glacial period that peaked 21,500 years ago was only the most recent of five glacial periods in the last 450,000 years. In fact, the Earth system has alternated between glacial and interglacial regimes for more than two million years, a period of time known as the Pleistocene. The duration and severity of the glacial periods increased during this period, with a particularly sharp change occurring between 900,000 and 600,000 years ago. Earth is currently within the most recent interglacial period, which started 11,700 years ago and is commonly known as the Holocene Epoch.

The continental glaciations of the Pleistocene left signatures on the landscape in the form of glacial deposits and landforms; however, the best knowledge of the magnitude and timing of the various glacial and interglacial periods comes from oxygen isotope records in ocean sediments. These records provide both a direct measure of sea level and an indirect measure of global ice volume. Water molecules composed of a lighter isotope of oxygen, ^{16}O, are evaporated more readily than molecules bearing a heavier isotope, ^{18}O. Glacial periods are characterized by high ^{18}O concentrations and represent a net transfer of water, especially with ^{16}O, from the oceans to the ice sheets. Oxygen isotope records indicate that interglacial periods have typically lasted 10,000–15,000 years, and maximum glacial periods were of similar length. Most of the past 500,000 years—approximately 80 percent—have been spent within various intermediate glacial states that were warmer than glacial maxima but cooler than interglacials. During these intermediate times, substantial glaciers occurred over much of Canada and probably covered Scandinavia as well. These intermediate states were not constant; they were characterized by continual, millennial-scale climate variation. There has been no average or typical state for global climate during Pleistocene and Holocene times; the Earth system has been in continual flux between interglacial and glacial patterns.

The cycling of the Earth system between glacial and interglacial modes has been ultimately driven by orbital variations. However, orbital forcing is by itself insufficient to explain all of this variation, and Earth system scientists are focusing their attention on the interactions and feedbacks between the myriad components of the Earth system. For example, the initial development of a continental ice sheet increases albedo over a portion of Earth, reducing surface absorption of sunlight and leading to further cooling. Similarly, changes in terrestrial vegetation, such as the replacement of forests by tundra, feed back into the atmosphere via changes in both albedo and latent heat flux from evapotranspiration. Forests—particularly those of tropical and temperate areas, with their large leaf area—release great amounts of water vapour and latent heat through transpiration. Tundra plants, which are much smaller, possess tiny leaves designed to slow water loss; they release only a small fraction of the water vapour that forests do.

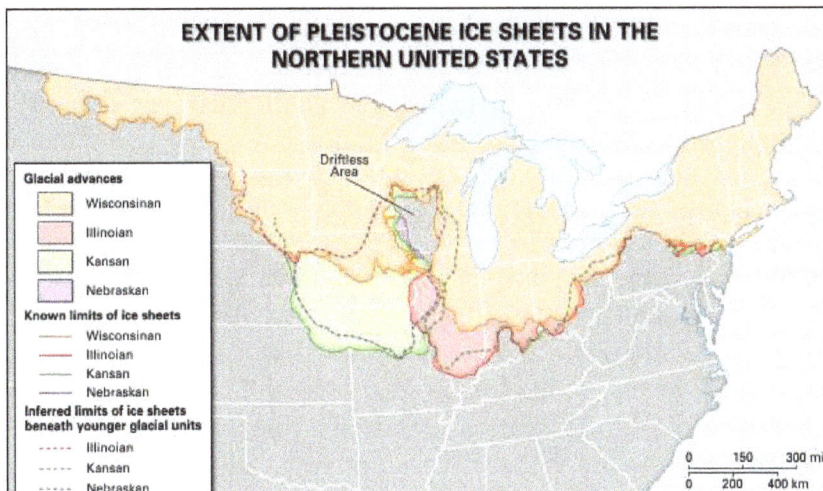

EXTENT OF PLEISTOCENE ICE SHEETS IN THE NORTHERN UNITED STATES

The discovery in ice core records that atmospheric concentrations of two potent greenhouse gases, carbon dioxide and methane, have decreased during past glacial periods and peaked during interglacials indicates important feedback processes in the Earth system. Reduction of greenhouse gas concentrations during the transition to a glacial phase would reinforce and amplify cooling already under way. The reverse is true for transition to interglacial periods. The glacial carbon sink remains a topic of considerable research activity. A full understanding of glacial-interglacial carbon dynamics requires knowledge of the complex interplay among ocean chemistry and circulation, ecology of marine and terrestrial organisms, ice sheet dynamics, and atmospheric chemistry and circulation.

The blue areas are those that were covered by ice sheets in the past. The Kansan and Nebraskan sheets overlapped almost the same areas, and the Wisconsin and Illinoisan sheets covered approximately the same territory. In the high altitudes of the West are the Cordilleran ice sheets. An area at the junction of Wisconsin, Minnesota, Iowa, and Illinois was never entirely covered with ice. Successive ice caps reached limits that differed only slightly. The area covered by ice at any time is shown in white.

The Last Great Cooling

The Earth system has undergone a general cooling trend for the past 50 million years, culminating in the development of permanent ice sheets in the Northern Hemisphere about 2.75 million years ago. These ice sheets expanded and contracted in a regular rhythm, with each glacial maximum separated from adjacent ones by 41,000 years (based on the cycle of axial tilt). As the ice sheets waxed and waned, global climate drifted steadily toward cooler conditions characterized by increasingly severe glaciations and increasingly cool interglacial phases. Beginning around 900,000 years ago, the glacial-interglacial cycles shifted frequency. Ever since, the glacial peaks have been 100,000 years apart, and the Earth system has spent more time in cool phases than before. The 41,000-year periodicity has continued, with smaller fluctuations superimposed on the 100,000-year cycle. In addition, a smaller, 23,000-year cycle has occurred through both the 41,000-year and 100,000-year cycles.

The 23,000-year and 41,000-year cycles are driven ultimately by two components of Earth's orbital geometry: the equinoctial precession cycle (23,000 years) and the axial-tilt cycle (41,000 years). Although the third parameter of Earth's orbit, eccentricity, varies on a 100,000-year cycle, its magnitude is insufficient to explain the 100,000-year cycles of glacial and interglacial periods of the past 900,000 years. The origin of the periodicity present in Earth's eccentricity is an important question in current paleoclimate research.

Climate Change through Geologic Time

The Earth system has undergone dramatic changes throughout its 4.5-billion-year history. These have included climatic changes diverse in mechanisms, magnitudes, rates, and consequences. Many of these past changes are obscure and controversial, and some have been discovered only recently. Nevertheless, the history of life has been strongly influenced by these changes, some of which radically altered the course of evolution. Life itself is implicated as a causative agent of some of these changes, as the processes of photosynthesis and respiration have largely shaped the chemistry of Earth's atmosphere, oceans, and sediments.

Cenozoic Climates

The Cenozoic Era—encompassing the past 65.5 million years, the time that has elapsed since the mass extinction event marking the end of the Cretaceous Period—has a broad range of climatic variation characterized by alternating intervals of global warming and cooling. Earth has experienced both extreme warmth and extreme cold during this period. These changes have been driven by tectonic forces, which have altered the positions and elevations of the continents as well as ocean passages and bathymetry. Feedbacks between different components of the Earth system (atmosphere, biosphere, lithosphere, cryosphere, and oceans in the hydrosphere) are being increasingly recognized as influences of global and regional climate. In particular, atmospheric concentrations of carbon dioxide have varied substantially during the Cenozoic for reasons that are poorly understood, though its fluctuation must have involved feedbacks between Earth's spheres.

Orbital forcing is also evident in the Cenozoic, although, when compared on such a vast era-level timescale, orbital variations can be seen as oscillations against a slowly changing backdrop of lower-frequency climatic trends. Descriptions of the orbital variations have evolved according to the growing understanding of tectonic and biogeochemical changes. A pattern emerging from recent paleoclimatologic studies suggests that the climatic effects of eccentricity, precession, and axial tilt have been amplified during cool phases of the Cenozoic, whereas they have been dampened during warm phases.

The meteor impact that occurred at or very close to the end of the Cretaceous came at a time of global warming, which continued into the early Cenozoic. Tropical and subtropical flora and fauna occurred at high latitudes until at least 40 million years ago, and geochemical records of marine sediments have indicated the presence of warm oceans. The interval of maximum temperature occurred during the late Paleocene and early Eocene epochs (58.7 million to 40.4 million years ago). The highest global temperatures of the Cenozoic occurred during the Paleocene-Eocene Thermal Maximum (PETM), a short interval lasting approximately 100,000 years. Although the underlying causes are unclear, the onset of the PETM about 56 million years ago was rapid, occurring within a few thousand years, and ecological consequences were large, with widespread extinctions in both marine and terrestrial ecosystems. Sea surface and continental air temperatures increased by more than 5 °C (9 °F) during the transition into the PETM. Sea surface temperatures in the high-latitude Arctic may have been as warm as 23 °C (73 °F), comparable to modern subtropical and warm-temperate seas. Following the PETM, global temperatures declined to pre-PETM levels, but they gradually increased to near-PETM levels over the next few million years during a period known as the Eocene Optimum. This temperature maximum was followed by a steady decline in global temperatures toward the Eocene-Oligocene boundary, which occurred about 33.9 million years ago. These changes are well-represented in marine sediments and in paleontological records from the continents, where vegetation zones moved Equator-ward. Mechanisms underlying the cooling trend are under study, but it is most likely that tectonic movements played an important role. This period saw the gradual opening of the sea passage between Tasmania and Antarctica, followed by the opening of the Drake Passage between South America and Antarctica. The latter, which isolated Antarctica within a cold polar sea, produced global effects on atmospheric and oceanic circulation. Recent evidence suggests that decreasing atmospheric concentrations of carbon dioxide during this period may have initiated a steady and irreversible cooling trend over the next few million years.

A continental ice sheet developed in Antarctica during the Oligocene Epoch, persisting until a rapid warming event took place 27 million years ago. The late Oligocene and early to mid-Miocene epochs (28.4 million to 13.8 million years ago) were relatively warm, though not nearly as warm as the Eocene. Cooling resumed 15 million years ago, and the Antarctic Ice Sheet expanded again to cover much of the continent. The cooling trend continued through the late Miocene and accelerated into the early Pliocene Epoch, 5.3 million years ago. During this period the Northern Hemisphere remained ice-free, and paleobotanical studies show cool-temperate Pliocene floras at high latitudes on Greenland and the Arctic Archipelago. The Northern Hemisphere glaciation, which began 3.2 million years ago, was driven by tectonic events, such as the closing of the Panama seaway and the uplift of the Andes, the Tibetan Plateau, and western parts of North America. These tectonic events led to changes in the circulation of the oceans and the atmosphere, which in turn fostered the development of persistent ice at high northern latitudes. Small-magnitude variations in carbon dioxide concentrations, which had been relatively low since at least the mid-Oligocene (28.4 million years ago), are also thought to have contributed to this glaciation.

Phanerozoic Climates

The Phanerozoic Eon (542 million years ago to the present), which includes the entire span of complex, multicellular life on Earth, has witnessed an extraordinary array of climatic states and transitions. The sheer antiquity of many of these regimes and events renders them difficult to understand in detail. However, a number of periods and transitions are well known, owing to good geological records and intense study by scientists. Furthermore, a coherent pattern of low-frequency climatic variation is emerging, in which the Earth system alternates between warm ("greenhouse") phases and cool ("icehouse") phases. The warm phases are characterized by high temperatures, high sea levels, and an absence of continental glaciers. Cool phases in turn are marked by low temperatures, low sea levels, and the presence of continental ice sheets, at least at high latitudes. Superimposed on these alternations are higher-frequency variations, where cool periods are embedded within greenhouse phases and warm periods are embedded within icehouse phases. For example, glaciers developed for a brief period (between 1 million and 10 million years) during the late Ordovician and early Silurian, in the middle of the early Paleozoic greenhouse phase (542 million to 350 million years ago). Similarly, warm periods with glacial retreat occurred within the late Cenozoic cool period during the late Oligocene and early Miocene epochs.

The Earth system has been in an icehouse phase for the past 30 million to 35 million years, ever since the development of ice sheets on Antarctica. The previous major icehouse phase occurred between about 350 million and 250 million years ago, during the Carboniferous and Permian periods of the late Paleozoic Era. Glacial sediments dating to this period have been identified in much of Africa as well as in the Arabian Peninsula, South America, Australia, India, and Antarctica. At the time, all these regions were part of Gondwana, a high-latitude supercontinent in the Southern Hemisphere. The glaciers atop Gondwana extended to at least 45° S latitude, similar to the latitude reached by Northern Hemisphere ice sheets during the Pleistocene. Some late Paleozoic glaciers extended even further Equator-ward—to 35° S. One of the most striking features of this time period are cyclothems, repeating sedimentary beds of alternating sandstone, shale, coal, and limestone. The great coal deposits of North America's Appalachian region, the American Midwest, and northern Europe are interbedded in these cyclothems, which may represent repeated transgressions (producing limestone) and retreats (producing shales and coals) of ocean shorelines in response to orbital variations.

The two most prominent warm phases in Earth history occurred during the Mesozoic and early Cenozoic eras (approximately 250 million to 35 million years ago) and the early and mid-Paleozoic (approximately 500 million to 350 million years ago). Climates of each of these greenhouse periods were distinct; continental positions and ocean bathymetry were very different, and terrestrial vegetation was absent from the continents until relatively late in the Paleozoic warm period. Both of these periods experienced substantial long-term climate variation and change; increasing evidence indicates brief glacial episodes during the mid-Mesozoic.

Understanding the mechanisms underlying icehouse-greenhouse dynamics is an important area of research, involving an interchange between geologic records and the modeling of the Earth system and its components. Two processes have been implicated as drivers of Phanerozoic climate change. First, tectonic forces caused changes in the positions and elevations of continents and the bathymetry of oceans and seas. Second, variations in greenhouse gases were also important drivers of climate, though at these long timescales they were largely controlled by tectonic processes, in which sinks and sources of greenhouse gases varied.

Climates of Early Earth

The pre-Phanerozoic interval, also known as Precambrian time, comprises some 88 percent of the time elapsed since the origin of Earth. The pre-Phanerozoic is a poorly understood phase of Earth system history. Much of the sedimentary record of the atmosphere, oceans, biota, and crust of the early Earth has been obliterated by erosion, metamorphosis, and subduction. However, a number of pre-Phanerozoic records have been found in various parts of the world, mainly from the later portions of the period. Pre-Phanerozoic Earth system history is an extremely active area of research, in part because of its importance in understanding the origin and early evolution of life on Earth. Furthermore, the chemical composition of Earth's atmosphere and oceans largely developed during this period, with living organisms playing an active role. Geologists, paleontologists, microbiologists, planetary geologists, atmospheric scientists, and geochemists are focusing intense efforts on understanding this period. Three areas of particular interest and debate are the "faint young Sun paradox," the role of organisms in shaping Earth's atmosphere, and the possibility that Earth went through one or more "snowball" phases of global glaciation.

Faint Young Sun Paradox

Astrophysical studies indicate that the luminosity of the Sun was much lower during Earth's early history than it has been in the Phanerozoic. In fact, radiative output was low enough to suggest that all surface water on Earth should have been frozen solid during its early history, but evidence shows that it was not. The solution to this "faint young Sun paradox" appears to lie in the presence of unusually high concentrations of greenhouse gases at the time, particularly methane and carbon dioxide. As solar luminosity gradually increased through time, concentrations of greenhouse gases would have to have been much higher than today. This circumstance would have caused Earth to heat up beyond life-sustaining levels. Therefore, greenhouse gas concentrations must have decreased proportionally with increasing solar radiation, implying a feedback mechanism to regulate greenhouse gases. One of these mechanisms might have been rock weathering, which is temperature-dependent and serves as an important sink for, rather than source of, carbon dioxide by removing sizable amounts of this gas from the atmosphere. Scientists are also looking to biological processes (many of which also serve as carbon dioxide sinks) as complementary or alternative regulating mechanisms of greenhouse gases on the young Earth.

Photosynthesis and Atmospheric Chemistry

The evolution by photosynthetic bacteria of a new photosynthetic pathway, substituting water (H_2O) for hydrogen sulfide (H_2S) as a reducing agent for carbon dioxide, had dramatic consequences for Earth system geochemistry. Molecular oxygen (O_2) is given off as a by-product of photosynthesis using the H_2O pathway, which is energetically more efficient than the more primitive H_2S pathway. Using H_2O as a reducing agent in this process led to the large-scale deposition of banded-iron formations, or BIFs, a source of 90 percent of present-day iron ores. Oxygen present in ancient oceans oxidized dissolved iron, which precipitated out of solution onto the ocean floors. This deposition process, in which oxygen was used up as fast as it was produced, continued for millions of years until most of the iron dissolved in the oceans was precipitated. By approximately 2 billion years ago, oxygen was able to accumulate in dissolved form in seawater and to outgas to the atmosphere. Although oxygen does not have greenhouse gas properties, it plays important indirect roles in Earth's climate, particularly in phases of the carbon cycle. Scientists are studying the role of oxygen and other contributions of early life to the development of the Earth system.

Snowball Earth Hypothesis

Geochemical and sedimentary evidence indicates that Earth experienced as many as four extreme cooling events between 750 million and 580 million years ago. Geologists have proposed that Earth's oceans and land surfaces were covered by ice from the poles to the Equator during these events. This "Snowball Earth" hypothesis is a subject of intense study and discussion. Two important questions arise from this hypothesis. First, how, once frozen, could Earth thaw? Second, how could life survive periods of global freezing? A proposed solution to the first question involves the outgassing of massive amounts of carbon dioxide by volcanoes, which could have warmed the planetary surface rapidly, especially given that major carbon dioxide sinks (rock weathering and photosynthesis) would have been dampened by a frozen Earth. A possible answer to the second question may lie in the existence of present-day life-forms within hot springs and deep-sea vents, which would have persisted long ago despite the frozen state of Earth's surface.

A counter-premise known as the "Slushball Earth" hypothesis contends that Earth was not completely frozen over. Rather, in addition to massive ice sheets covering the continents, parts of the planet (especially ocean areas near the Equator) could have been draped only by a thin, watery layer of ice amid areas of open sea. Under this scenario, photosynthetic organisms in low-ice or ice-free regions could continue to capture sunlight efficiently and survive these periods of extreme cold.

Prior to the 18th century, scientists had not suspected that prehistoric climates were different from the modern period. By the late 18th century, geologists found evidence of a succession of geological ages with changes in climate. In the years since, a great deal of scientific progress has been made understanding the workings of the climate system.

Causes

On the broadest scale, the rate at which energy is received from the Sun and the rate at which it is lost to space determine the equilibrium temperature and climate of Earth. This energy is distributed around the globe by winds, ocean currents, and other mechanisms to affect the climates of different regions.

Factors that can shape climate are called climate forcings or "forcing mechanisms". These include processes such as variations in solar radiation, variations in the Earth's orbit, variations in the albedo or reflectivity of the continents, atmosphere, and oceans, mountain-building and continental drift and changes in greenhouse gas concentrations. There are a variety of climate change feedbacks that can either amplify or diminish the initial forcing. Some parts of the climate system, such as the oceans and ice caps, respond more slowly in reaction to climate forcings, while others respond more quickly. There are also key threshold factors which when exceeded can produce rapid change.

Forcing mechanisms can be either "internal" or "external". Internal forcing mechanisms are natural processes within the climate system itself (e.g., the thermohaline circulation). External forcing mechanisms can be either anthropogenic (e.g. increased emissions of greenhouse gases and dust) or natural (e.g., changes in solar output, the earth's orbit, volcano eruptions).

Whether the initial forcing mechanism is internal or external, the response of the climate system might be fast (e.g., a sudden cooling due to airborne volcanic ash reflecting sunlight), slow (e.g. thermal expansion of warming ocean water), or a combination (e.g., sudden loss of albedo in the Arctic Ocean as sea ice melts, followed by more gradual thermal expansion of the water). Therefore, the climate system can respond abruptly, but the full response to forcing mechanisms might not be fully developed for centuries or even longer.

Climate Variability

Scientists generally define the five components of earth's climate system to include atmosphere, hydrosphere, cryosphere, lithosphere (restricted to the surface soils, rocks, and sediments), and biosphere. Natural changes in the climate system result in internal "climate variability". Examples include the type and distribution of species, and changes in ocean-atmosphere circulations.

Ocean-atmosphere Variability

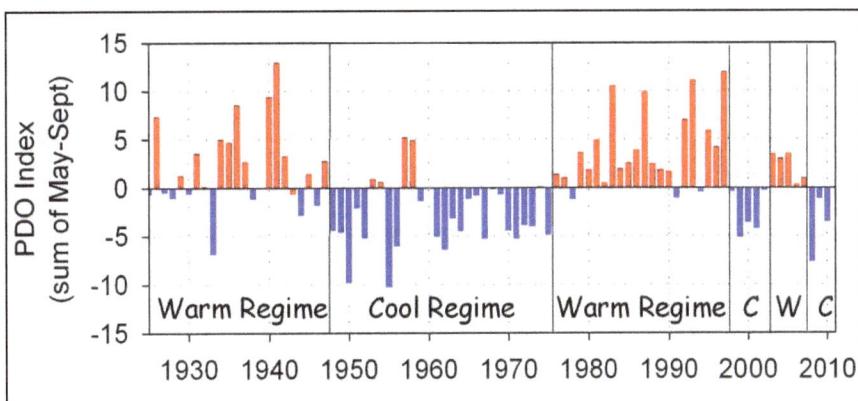

Pacific decadal oscillation 1925 to 2010.

The ocean and atmosphere can work together to spontaneously generate internal climate variability that can persist for years to decades at a time. Examples of this type of variability include the El Niño–Southern Oscillation, the Pacific decadal oscillation, and the Atlantic Multidecadal Oscillation. These variations can affect global average surface temperature by redistributing heat between the deep ocean and the atmosphere and/or by altering the cloud/water vapor/sea ice distribution which can affect the total energy budget of the earth.

The oceanic aspects of these circulations can generate variability on centennial timescales due to the ocean having hundreds of times more mass than in the atmosphere, and thus very high thermal inertia. For example, alterations to ocean processes such as thermohaline circulation play a key role in redistributing heat in the world's oceans. Due to the long timescales of this circulation, ocean temperature at depth is still adjusting to effects of the Little Ice Age which occurred between the 1600 and 1800s.

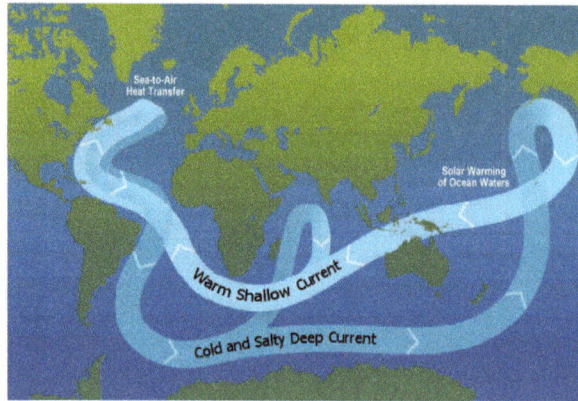

A schematic of modern thermohaline circulation. Tens of millions of years ago, continental-plate movement formed a land-free gap around Antarctica, allowing the formation of the ACC, which keeps warm waters away from Antarctica.

Life

Life affects climate through its role in the carbon and water cycles and through such mechanisms as albedo, evapotranspiration, cloud formation, and weathering. Examples of how life may have affected past climate include:

- Glaciation 2.3 billion years ago triggered by the evolution of oxygenic photosynthesis, which depleted the atmosphere of the greenhouse gas carbon dioxide and introduced free oxygen,

- Another glaciation 300 million years ago ushered in by long-term burial of decomposition-resistant detritus of vascular land-plants (creating a carbon sink and forming coal),

- Termination of the Paleocene–Eocene Thermal Maximum 55 million years ago by flourishing marine phytoplankton,

- Reversal of global warming 49 million years ago by 800,000 years of arctic azolla blooms,

- Global cooling over the past 40 million years driven by the expansion of grass-grazer ecosystems.

External Forcing Mechanisms

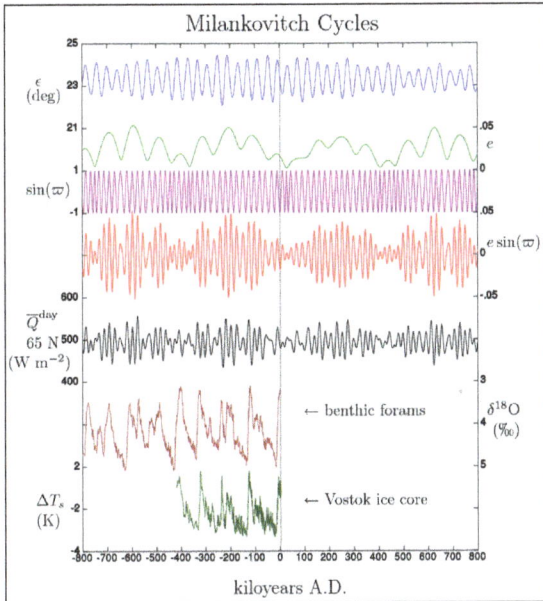

Milankovitch cycles from 800,000 years ago in the past to 800,000 years in the future.

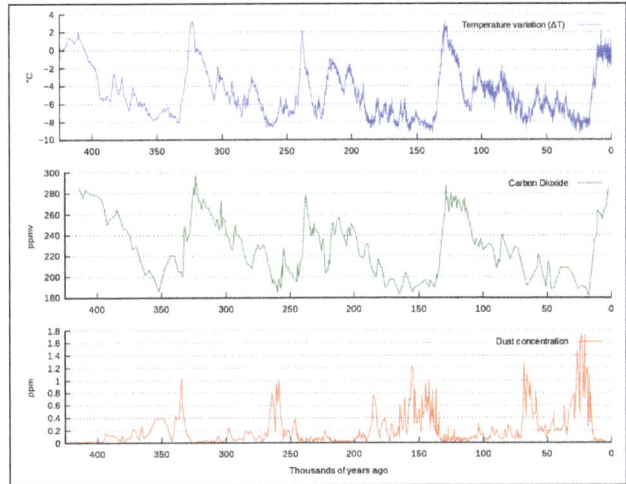

Variations in CO_2, temperature and dust from the Vostok ice core over the last 450,000 years.

Human Influences

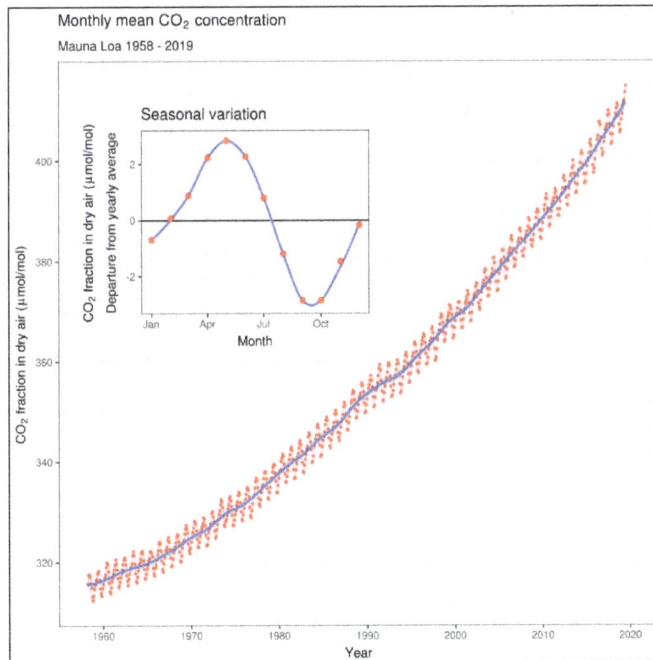

Increase in atmospheric CO_2 levels.

In the context of climate variation, anthropogenic factors are human activities which affect the climate. The scientific consensus on climate change is "that climate is changing and that these changes are in large part caused by human activities", and it "is largely irreversible".

"There is a strong, credible body of evidence, based on multiple lines of research, documenting that climate is changing and that these changes are in large part caused by human activities. While much remains to be learned, the core phenomenon, scientific questions, and hypotheses have been examined thoroughly and have stood firm in the face of serious scientific debate and careful evaluation of alternative explanations."

— *United States National Research Council, Advancing the Science of Climate Change*

Of most concern in these anthropogenic factors is the increase in CO_2 levels. This is due to emissions from fossil fuel combustion, followed by aerosols (particulate matter in the atmosphere), and the CO_2 released by cement manufacture. Other factors, including land use, ozone depletion, animal husbandry (ruminant animals such as cattle produce methane, as do termites), and deforestation, are also of concern in the roles they play—both separately and in conjunction with other factors—in affecting climate, microclimate, and measures of climate variables.

Orbital Variations

Slight variations in Earth's motion lead to changes in the seasonal distribution of sunlight reaching the Earth's surface and how it is distributed across the globe. There is very little change to the area-averaged annually averaged sunshine; but there can be strong changes in the geographical and seasonal distribution. The three types of kinematic change are variations in Earth's eccentricity, changes in the tilt angle of Earth's axis of rotation, and precession of Earth's axis. Combined together, these produce Milankovitch cycles which affect climate and are notable for their correlation to glacial and interglacial periods, their correlation with the advance and retreat of the Sahara, and for their appearance in the stratigraphic record.

The IPCC notes that Milankovitch cycles drove the ice age cycles, CO_2 followed temperature change "with a lag of some hundreds of years", and that as a feedback amplified temperature change. The depths of the ocean have a lag time in changing temperature (thermal inertia on such scale). Upon seawater temperature change, the solubility of CO_2 in the oceans changed, as well as other factors affecting air-sea CO_2 exchange.

Solar Output

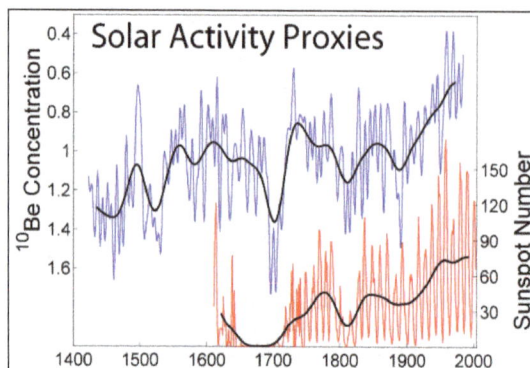

Variations in solar activity during the last several centuries based on observations of sunspots and beryllium isotopes. The period of extraordinarily few sunspots in the late 17th century was the Maunder minimum.

The Sun is the predominant source of energy input to the Earth. Other sources include geothermal energy from the Earth's core, tidal energy from the Moon and heat from the decay of radioactive compounds. Both long- and short-term variations in solar intensity are known to affect global climate.

Three to four billion years ago, the Sun emitted only 75% as much power as it does today. If the atmospheric composition had been the same as today, liquid water should not have existed on the Earth's surface. However, there is evidence for the presence of water on the early Earth, in the Hadean and Archean eons, leading to what is known as the faint young Sun paradox. Hypothesized solutions to this paradox include a vastly different atmosphere, with much higher concentrations of greenhouse gases than currently exist. Over the following approximately 4 billion years, the energy output of the Sun increased and atmospheric composition changed. The Great Oxygenation Event—oxygenation of the atmosphere around 2.4 billion years ago—was the most notable alteration. Over the next five billion years from the present, the Sun's ultimate death as it becomes a red giant and then a white dwarf will have large effects on climate, with the red giant phase possibly ending any life on Earth that survives until that time.

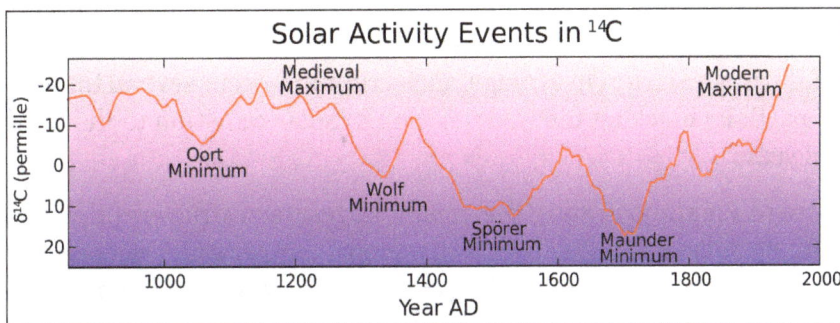

Solar activity events recorded in radiocarbon. Values since 1950 not shown.

Solar output varies on shorter time scales, including the 11-year solar cycle and longer-term modulations. Solar intensity variations, possibly as a result of the Wolf, Spörer, and the Maunder Minima, are considered to have been influential in triggering the Little Ice Age. This event extended from 1550 to 1850 AD and was marked by relative cooling and greater glacier extent than the centuries before and afterward. Solar variation may also have affected some of the warming observed from 1900 to 1950. The cyclical nature of the Sun's energy output is not yet fully understood; it differs from the very slow change that is happening within the Sun as it ages and evolves.

Some studies point toward solar radiation increases from cyclical sunspot activity affecting global warming, and climate may be influenced by the sum of all effects (solar variation, anthropogenic radiative forcings, etc.).

A 2010 study suggests "that the effects of solar variability on temperature throughout the atmosphere may be contrary to current expectations".

In 2011, CERN announced the initial results from its CLOUD experiment in the *Nature* journal. The results indicate that ionisation from cosmic rays significantly enhances aerosol formation in the presence of sulfuric acid and water, but in the lower atmosphere where ammonia is also required, this is insufficient to account for aerosol formation and additional trace vapours must be involved. The next step is to find more about these trace vapours, including whether they are of natural or human origin.

Volcanism

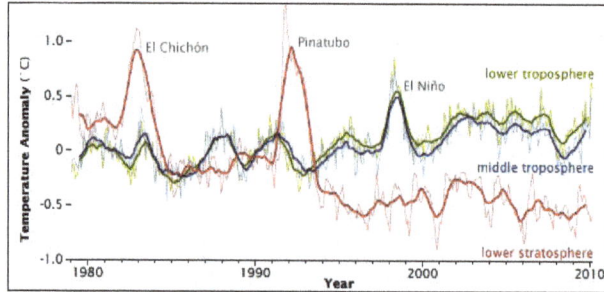

In atmospheric temperature from 1979 to 2010, determined by MSU NASA satellites, effects appear from aerosols released by major volcanic eruptions (El Chichón and Pinatubo). El Niño is a separate event, from ocean variability.

The eruptions considered to be large enough to affect the Earth's climate on a scale of more than 1 year are the ones that inject over 100,000 tons of SO_2 into the stratosphere. This is due to the optical properties of SO_2 and sulfate aerosols, which strongly absorb or scatter solar radiation, creating a global layer of sulfuric acid haze. On average, such eruptions occur several times per century, and cause cooling (by partially blocking the transmission of solar radiation to the Earth's surface) for a period of several years.

The eruption of Mount Pinatubo in 1991, the second largest terrestrial eruption of the 20th century, affected the climate substantially, subsequently global temperatures decreased by about 0.5 °C (0.9 °F) for up to three years. Thus, the cooling over large parts of the Earth reduced surface temperatures in 1991–93, the equivalent to a reduction in net radiation of 4 watts per square meter. The Mount Tambora eruption in 1815 caused the Year Without a Summer. Much larger eruptions, known as large igneous provinces, occur only a few times every 50,000,000–100,000,000 years through flood basalts; in Earth's past, they have caused global warming and mass extinctions.

Small eruptions, with injections of less than 0.1 Mt of sulfur dioxide into the stratosphere, affect the atmosphere only subtly, as temperature changes are comparable with natural variability. However, because smaller eruptions occur at a much higher frequency, they too significantly affect Earth's atmosphere.

Seismic monitoring maps current and future trends in volcanic activities, and tries to develop early warning systems. In climate modelling the aim is to study the physical mechanisms and feedbacks of volcanic forcing.

Volcanoes are also part of the extended carbon cycle. Over very long (geological) time periods, they release carbon dioxide from the Earth's crust and mantle, counteracting the uptake by sedimentary rocks and other geological carbon dioxide sinks. The US Geological Survey estimates are that volcanic emissions are at a much lower level than the effects of current human activities, which generate 100–300 times the amount of carbon dioxide emitted by volcanoes. A review of published studies indicates that annual volcanic emissions of carbon dioxide, including amounts released from mid-ocean ridges, volcanic arcs, and hot spot volcanoes, are only the equivalent of 3 to 5 days of human-caused output. The annual amount put out by human activities may be greater than the amount released by supereruptions, the most recent of which was the Toba eruption in Indonesia 74,000 years ago.

Although volcanoes are technically part of the lithosphere, which itself is part of the climate system, the IPCC explicitly defines volcanism as an external forcing agent.

Plate Tectonics

Over the course of millions of years, the motion of tectonic plates reconfigures global land and ocean areas and generates topography. This can affect both global and local patterns of climate and atmosphere-ocean circulation.

The position of the continents determines the geometry of the oceans and therefore influences patterns of ocean circulation. The locations of the seas are important in controlling the transfer of heat and moisture across the globe, and therefore, in determining global climate. A recent example of tectonic control on ocean circulation is the formation of the Isthmus of Panama about 5 million years ago, which shut off direct mixing between the Atlantic and Pacific Oceans. This strongly affected the ocean dynamics of what is now the Gulf Stream and may have led to Northern Hemisphere ice cover. During the Carboniferous period, about 300 to 360 million years ago, plate tectonics may have triggered large-scale storage of carbon and increased glaciation. Geologic evidence points to a "megamonsoonal" circulation pattern during the time of the supercontinent Pangaea, and climate modeling suggests that the existence of the supercontinent was conducive to the establishment of monsoons.

The size of continents is also important. Because of the stabilizing effect of the oceans on temperature, yearly temperature variations are generally lower in coastal areas than they are inland. A larger supercontinent will therefore have more area in which climate is strongly seasonal than will several smaller continents or islands.

Other Mechanisms

The Earth receives an influx of ionized particles known as cosmic rays from a variety of external sources, including the Sun. A hypothesis holds that an increase in the cosmic ray flux would increase the ionization in the atmosphere, leading to greater cloud cover. This, in turn, would tend to cool the surface. The non-solar cosmic ray flux may vary as a result of a nearby supernova event, the solar system passing through a dense interstellar cloud, or the oscillatory movement of the Sun's position with respect to the galactic plane. The latter can increase the flux of high-energy cosmic rays coming from the Virgo cluster.

Evidence exists that the Chicxulub impact some 66 million years ago had severely affected the Earth's climate. Large quantities of sulfate aerosols were kicked up into the atmosphere, decreasing global temperatures by up to 26 °C and producing sub-freezing temperatures for a period of 3–16 years. The recovery time for this event took more than 30 years.

Climate Sensitivity

Climate sensitivity is the globally averaged temperature change in response to changes in radiative forcing, which can occur, for instance, due to increased levels of carbon dioxide (CO_2). Although the term climate sensitivity is usually used in the context of radiative forcing by CO_2, it is thought of as a general property of the climate system: The change in surface air temperature following a unit change in radiative forcing, and the climate sensitivity parameter is therefore expressed in units of °C/(W/m²). For this to be useful, the measure must be independent of the nature of the forcing (e.g. from greenhouse gases or solar variation); which is true approximately. When climate sensitivity is expressed for a doubling of CO_2, its units are degrees Celsius (°C).

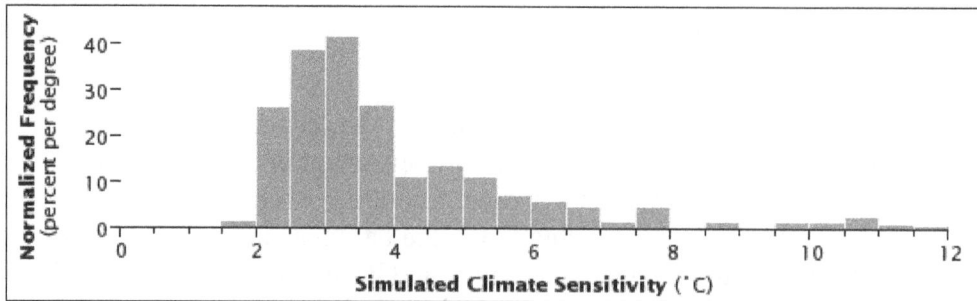

Frequency distribution of equilibrium climate sensitivity, based on model simulations.

Each model simulation has a different guess at processes that scientist don't understand sufficiently well. Few of the simulations result in less than 2 °C of warming or significantly more than 4 °C. However the positive skew, which is also found in other studies, suggests that if carbon dioxide concentrations double, the probability of very large increases in temperature is greater than the probability of very small increases.

In the context of global warming, different measures of climate sensitivity are used. The *equilibrium climate sensitivity* (ECS) is the temperature increase in °C that would result from sustained doubling of the concentration of carbon dioxide in Earth's atmosphere, after the Earth's energy budget and the climate system reach radiative equilibrium. The *transient climate response* (TCR) is the amount of temperature increase that might occur at the time when CO_2 doubles, having increased gradually by 1% each year (compounded). The *earth system sensitivity* (ESS) includes the effects of very-long-term Earth system feedback loops, such as changes in ice sheets or changes in the distribution of vegetative cover.

Climate sensitivity is typically estimated in three ways; by using observations taken during the industrial age, by using temperature and other data from the Earth's past and by modelling the climate system in computers. For coupled atmosphere-ocean global climate models the climate sensitivity is an emergent property; rather than being a model parameter it is a result of a combination of model physics and parameters. By contrast, simpler energy-balance models may have climate sensitivity as an explicit parameter.

Different forms of Climate Sensitivity

A component of climate sensitivity is directly due to radiative forcing, for instance by CO_2, and a further contribution arises from climate feedback, both positive and negative. Without feedbacks the radiative forcing of approximately 3.7 W/m², due to doubling CO_2 from the pre-industrial 280 ppm, would eventually result in roughly 1 °C global warming. This is easy to calculate and undisputed. The uncertainty is due entirely to feedbacks in the system: the water vapor feedback, the ice-albedo feedback, the cloud feedback, and the lapse rate feedback. Due to climate inertia, the climate sensitivity depends upon the timescale in which one is interested. The transient response is defined by scientists as the temperature response over human time scales of around 70 years, the equilibrium climate sensitivity over centuries, and finally the Earth system sensitivity after multiple millennia.

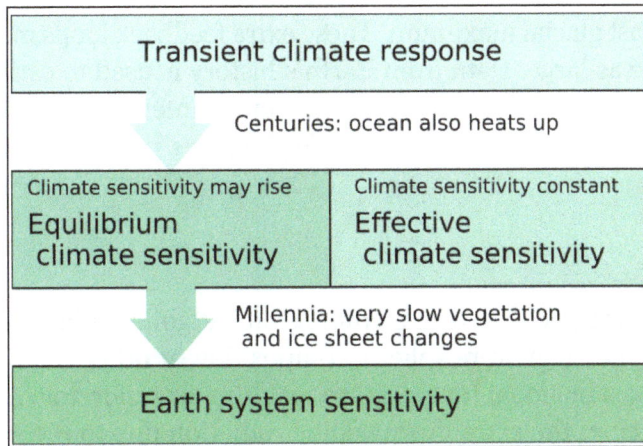

Schematic of how different measures of climate sensitivity relate to one another.

Equilibrium Climate Sensitivity

The equilibrium climate sensitivity (ECS) refers to the equilibrium change in global mean near-surface air temperature that would result from a sustained doubling of the atmospheric equivalent CO_2 concentration (ΔT_{2x}). A comprehensive model estimate of equilibrium sensitivity requires a very long model integration; fully equilibrating ocean temperatures requires the integration of thousands of model years, although it is possible to produce an estimate more quickly using the method of Gregory et al. As estimated by the IPCC Fifth Assessment Report (AR5), "there is high confidence that ECS is extremely unlikely less than 1 °C and medium confidence that the ECS is likely between 1.5 °C and 4.5 °C and very unlikely greater than 6 °C".

Effective Climate Sensitivity

The effective climate sensitivity is an estimate of equilibrium climate sensitivity using data from a climate system, either in a model or real-world observations, that is not yet in equilibrium. Estimation is done by using the assumption that the net effect of feedbacks as measured after a period of warming remains constant afterwards. This is not necessarily true, as feedbacks can change with time, or with the particular starting state or forcing history of the climate system.

Transient Climate Response

The transient climate response (TCR) is defined as the average temperature response over a twenty-year period centered at CO_2 doubling in a transient simulation with CO_2 increasing at 1% per year (compounded), i.e., 60 to 80 years following initiation of the increase in CO_2. The transient response is lower than the equilibrium sensitivity because the deep ocean, which takes many centuries to reach a new steady state after a perturbation, continues to serve as a sink for heat from the upper ocean. The IPCC literature assessment estimates that TCR likely lies between 1 °C and 2.5 °C. A related concept is the transient climate response to cumulative carbon emissions, which is the globally averaged surface temperature change per unit of CO_2 emitted.

Earth System Sensitivity

The Earth system sensitivity (ESS) includes the effects of slower feedback loops, such as the change in Earth's albedo from the melting of large ice sheets that covered much of the northern hemisphere during the last glacial maximum. These extra feedback loops make the ESS larger than the ECS – possibly twice as large. Data from Earth's history is used to estimate ESS, but climatic conditions were quite different which makes it difficult to infer information for future ESS. ESS includes the entire system except the carbon cycle. Changes in albedo as a result of vegetation changes are included.

Radiative Forcing

Radiative forcing is the imbalance between incoming and outgoing radiation at the top of the atmosphere resulting from a change in atmospheric composition or other changes in radiation budget prior to long-term changes in global temperature resulting from forcing. A number of inputs can give rise to radiative forcing; the extra downwelling radiation due to the greenhouse effect, solar radiation variability due to orbital changes, changes in solar irradiance, direct aerosol effects – for example changes in albedo due to cloud cover – indirect aerosol effects, and changes in land use.

Radiative forcing as a consequence of greenhouse gases is well understood while, as of 2013, large uncertainties remain for aerosols. In time-dependent estimates of climate sensitivity, the concept of the effective radiative forcing, which includes rapid adjustments in the stratosphere and the troposphere to the instantaneous radiative forcing, is usually used.

Sensitivity to Nature of the Forcing

Radiative forcing from sources other than CO_2 can cause a higher or lower surface warming than a similar radiative forcing due to CO_2; the amount of feedback varies, mainly because these forcings are not uniformly distributed over the globe. Forcings that initially warm the northern hemisphere, land, or polar regions more strongly; are systematically more effective at changing temperatures than an equivalent amount of CO_2 whose forcing is more uniformly distributed over the globe. Several studies indicate that aerosols are more effective than CO_2 at changing global temperatures while volcanic forcing is less effective. Ignoring these factors causes lower estimates of climate sensitivity when using radiative forcing and temperature records from the historical period.

State Dependence

While climate sensitivity is defined as the sensitivity to any doubling of CO_2, there is evidence that the sensitivity of the climate system is not always constant. In the absence of sea ice, for instance, a positive ice-albedo feedback loop is lost, making the system less sensitive overall. Because of this, the climate system may warm by a different amount after a first doubling of CO_2 than after a second doubling. The effect of this is small or negligible in the first century after CO_2 is released into the atmosphere. Furthermore, climate sensitivity may change if tipping points are crossed. It is unlikely that climate sensitivity increases instantly; rather, it changes at the time scale of the subsystem that is undergoing the tipping point. The more sensitive a climate system is to increased greenhouse gases, the more likely it is to have decades when temperatures are much higher or much lower than the longer-term average.

Estimating Climate Sensitivity

Climate sensitivity is often evaluated in terms of the change in equilibrium temperature due to radiative forcing caused by the greenhouse effect. The radiative forcing, and hence the change in temperature, is proportional to the logarithm of the concentration of infrared-absorbing ("greenhouse") gases in the atmosphere, as quantified by Arrhenius in the 19th century. The sensitivity of temperature to atmospheric gasses, most notably CO_2, is often expressed in terms of the change in temperature per doubling of the concentration of the gas.

Svante Arrhenius was the first person to quantify global warming as a consequence of a doubling of CO_2. In his first paper on the matter, he estimated that global temperature would rise by around 5 to 6 °C (9.0 to 10.8 °F) if the quantity of CO_2 was doubled. In later work he revised this estimate to 4 °C (7.2 °F). Arrhenius used the observations of radiation emitted by the full moon made by the astronomer Samuel Pierpont Langley to estimate the amount of radiation that was absorbed by water vapour and CO_2. He then assumed relative humidity would stay the same under global warming as a representation of the water vapour feedback.

A committee on anthropogenic global warming convened in 1979 by the National Academy of Sciences and chaired by Jule Charney estimated climate sensitivity to be 3 °C (5.4 °F) with a tolerance of 1.5 °C (2.7 °F). Only two sets of models were available; one, due to Syukuro Manabe, showed a climate sensitivity of 2 °C (3.6 °F) and the other, due to James E. Hansen, showed a climate sensitivity of 4 °C (7.2 °F). According to Manabe, "Charney chose 0.5 °C (0.90 °F) as a reasonable margin of error, subtracted it from Manabe's number, and added it to Hansen's, giving rise to the 1.5 to 4.5 °C (2.7 to 8.1 °F) range of likely climate sensitivity that has appeared in every greenhouse assessment."

In 2008 climatologist Stefan Rahmstorf wrote, regarding the Charney report's original range of uncertainty; "At that time, this range was on very shaky ground. Since then, many vastly improved models have been developed by a number of climate research centers around the world. Current state-of-the-art climate models span a range of 2.6 to 4.1 °C (4.7 to 7.4 °F), most clustering around 3 °C (5.4 °F)."

Intergovernmental Panel on Climate Change

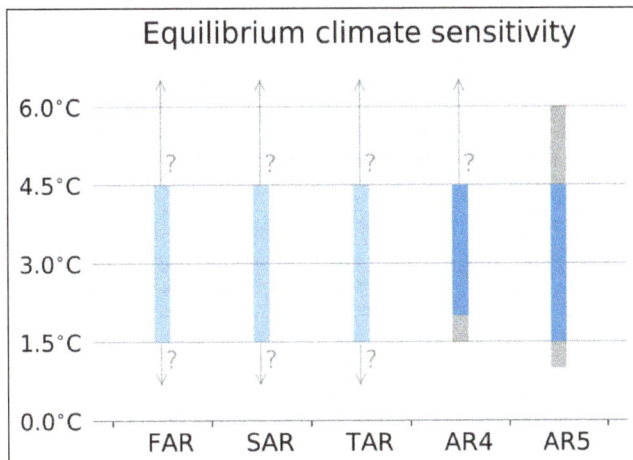

Historical estimates of climate sensitivity from the IPCC assessments. The first three reports gave a qualitative likely range, while the fourth and fifth assessment report formally quantified the uncertainty. The dark blue range is judged as being more than 66% likely.

After the publication of the Charney report, despite considerable progress in the understanding of the climate system, further assessments reported a similar range in climate sensitivity. The 1990 IPCC First Assessment Report estimated that equilibrium climate sensitivity to a doubling of CO_2 lay between 1.5 and 4.5 °C (2.7 and 8.1 °F), with a "best guess in the light of current knowledge" of 2.5 °C (4.5 °F). This report used models that had simplified representations of ocean dynamics. The IPCC supplementary report, 1992, which used full-ocean GCMs, saw "no compelling reason to warrant changing" from this estimate; and the IPCC Second Assessment Report said, "No strong reasons have emerged to change" these estimates. In these reports, much of the uncertainty was attributed to cloud processes. The 2001 IPCC TAR also retained this likely range.

Authors of the IPCC Fourth Assessment Report stated that confidence in estimates of equilibrium climate sensitivity had increased substantially since the Third Annual Report. IPCC authors concluded ECS is very likely to be greater than 1.5 °C (2.7 °F) and likely to lie in the range 2 to 4.5 °C (4 to 8.1 °F), with a most likely value of about 3 °C (5 °F). For fundamental physical reasons and data limitations, the IPCC stated a climate sensitivity higher than 4.5 °C (8.1 °F) cannot be ruled out but that agreement for these values with observations and "proxy" climate data is generally worse compared with values within the likely range.

The IPCC Fifth Assessment Report reverted to the earlier range of 1.5 to 4.5 °C (2.7 to 8.1 °F) (high confidence) because some estimates using industrial-age data came out low. They also stated that ECS is extremely unlikely to be less than 1 °C (1.8 °F) (high confidence), and is very unlikely to be greater than 6 °C (11 °F) (medium confidence). These values are estimated by combining the available data with expert judgement.

Using Industrial-age Data

Climate sensitivity can be estimated using observed temperature rise, observed ocean heat uptake, and modeled or observed radiative forcing. These data are linked though a simple energy-balance model to calculate climate sensitivity. Radiative forcing is often modeled, because satellites that measure it have not existed for the entire period. Estimates of climate sensitivity calculated from global energy constraints have consistently been lower than those calculated using other methods; estimates calculated using other methods have been around 2 °C (3.6 °F) or lower.

Estimates of transient climate response (TCR) calculated from models and observational data can be reconciled if it is taken into account that fewer temperature measurements are taken in the polar regions, which warm more quickly than average. If only regions for which measurements are available are used in evaluating the model, differences in TCR estimates almost disappear.

Rahmstorf provides an informal example of the estimation of climate sensitivity using observations made since the pre-industrial era, from which the following is modified. Denote the sensitivity, i.e. the equilibrium increase in global mean temperature including the effects of feedbacks due to a sustained forcing by doubled CO_2 ($F_{2\times CO_2}$; taken as 3.7 W/m²), as S °C. If Earth was to experience an equilibrium temperature change of ΔT (°C) due to a sustained forcing of ΔF (W/m²), then:

$$S = \Delta T \times F_{2\times CO_2} / \Delta F.$$

The global temperature increase since the beginning of the industrial period is about 0.8 °C (1.4

°F), and the radiative forcing due to CO_2 and other long-lived greenhouse gases – mainly methane, nitrous oxide, and chlorofluorocarbons – emitted since that time is about 2.6 W/m². Neglecting other forcings and considering the temperature increase to be an equilibrium increase would lead to a sensitivity of about 1.1 °C (2.0 °F). However, ΔF also contains contributions from solar activity (+0.3 W/m²), aerosols (−1 W/m²), ozone (0.3 W/m²), and other smaller influences, bringing the total forcing over the industrial period to 1.6 W/m² according to the best estimate of the IPCC AR4, with substantial uncertainty. The absence of equilibrium of the climate system must be accounted for by subtracting the planetary heat uptake rate H from the forcing; i.e.,

$$S = \Delta T \times F_{2 \times CO_2} / (\Delta F - H).$$

Taking planetary heat uptake rate as the rate of ocean heat uptake estimated by the IPCC AR4 as 0.2 W/m², yields a value for S of 2.1 °C (3.8 °F).

Other Strategies

The industrial-age temperatures can also be used to determine a timescale of the climate system, which is theoretically linked to climate sensitivity. If the effective heat capacity of the climate system is known and the timescale estimated by using the autocorrelation of the measured temperature, an estimate of climate sensitivity can be derived. Determination of both the timescale and heat capacity has proven difficult.

Attempts to use the 11-year solar cycle to constrain the transient climate response have been made. Solar irradiance is about 0.9 W/m² brighter during solar maximum than during solar minimum, which correlated with a variation of ±0.1 °C (0.18 °F) in measured average global temperature between the peak and minimum over the period 1959-2004. The solar minima in this period coincided with volcanic eruptions, which have a cooling effect on the global temperature. Because this causes a larger radiative forcing than the solar variations, it is questionable whether much information can be derived from the temperature variations.

Using Data from Earth's Past

Climate sensitivity can be estimated by using reconstructions of Earth's past temperatures and CO_2 levels. Different geological periods, for instance the warm Pliocene and the colder Pleistocene, are studied. Scientist seek periods that are in some sense analogous or informative to current climate change. As more information about the recent past becomes available, recent periods such as the Mid-holocene that occurred about 6,000 years ago and the Last Glacial Maximum (LGM) that took place about 21,000 years ago are often chosen.

The LGM was a period with a significantly lower global mean temperature than present day, and with a relatively well-known CO_2 concentration and radiative forcing in general. While orbital forcing was different from the present, this had little effect on mean annual temperatures. Different approaches to the task of estimating climate sensitivity from the LGM are taken. One approach is using estimates of global radiative forcing and temperature directly. The set of feedbacks active during the LGM, however, may be different than the feedbacks due to doubling CO_2, introducing additional uncertainty. In a different approach, a single model of intermediate complexity is run using a set of parameters so that each version has a different ECS. Those model versions that can best simulate the cooling during the LGM are thought to have the best ECS values. Others use an ensemble of different models.

Over the last 800,000 years, climate sensitivity has been found to be larger in cold periods than in warm periods. Climates further back in Earth's history are also used; an additional difficulty is that CO_2 concentrations cannot be readily obtained from ice cores so they must be estimated less directly. An estimate made using data from a major part of the Phanerozoic is consistent with sensitivities of current climate models and with other determinations. The Paleocene–Eocene Thermal Maximum provides a good opportunity to study the climate system when it is in a warm state.

Using Climate Models

Climate models of earth, for example the Coupled model intercomparison project (CMIP), are used to simulate the quantity of warming that will occur with rising CO_2 concentrations. The models are based on physical laws and represent the biosphere. Because of limited computer power, the physical laws have to be approximated, which leads to a wide range of estimates of climate sensitivity. Because the physical laws are programmed in a bottom-up way, climate sensitivity is an emergent property of these models.

Many CMIP phase 6 (CMIP6) models are being developed: some show climate sensitivity around 5 °C (9.0 °F), however the CMIP6 models have yet to be thoroughly independently analysed and researchers do not yet understand why some show this higher sensitivity.

Constrained Models

Bottom-up modelling of the climate system can lead to a wide range of outcomes. Models are often run using different plausible parameters in their approximation of physical laws and the behaviour of the biosphere; a so-called perturbed physics ensemble. Alternatively, structurally different models developed at different institutions are put together, creating an ensemble. By selecting only those simulations that can simulate some part of the historical climate well, a constrained estimate of climate sensitivity can be made. One strategy is the placing of more trust in climate models that perform well in general.

Alternatively, specific metrics that are directly and physically linked to climate sensitivity are sought; examples of this are the global patterns of warming, the ability of the models to reproduce observed relative humidity in the tropics and sub-tropics, patterns of radiation, and the variability of temperature about long term historical warming. When using ensemble climate models developed in different institutions, many of these constrained estimates of ECS are slightly higher than 3 °C (5.4 °F); the models with ECS slightly above 3 °C (5.4 °F) perform better in these metrics than models with a low climate sensitivity.

Economics

Because the economics of climate change mitigation depend a lot on how quickly carbon neutrality needs to be achieved, climate sensitivity is very important economically: one study suggests that halving the uncertainty of the transient climate response could save trillions of dollars.

Global Warming

Global Warming is the phenomenon of increasing average air temperatures near the surface of Earth over the past one to two centuries. Climate scientists have since the mid-20th century gathered detailed observations of various weather phenomena (such as temperatures, precipitation, and storms) and of related influences on climate (such as ocean currents and the atmosphere's chemical composition). These data indicate that Earth's climate has changed over almost every conceivable timescale since the beginning of geologic time and that the influence of human activities since at least the beginning of the Industrial Revolution has been deeply woven into the very fabric of climate change.

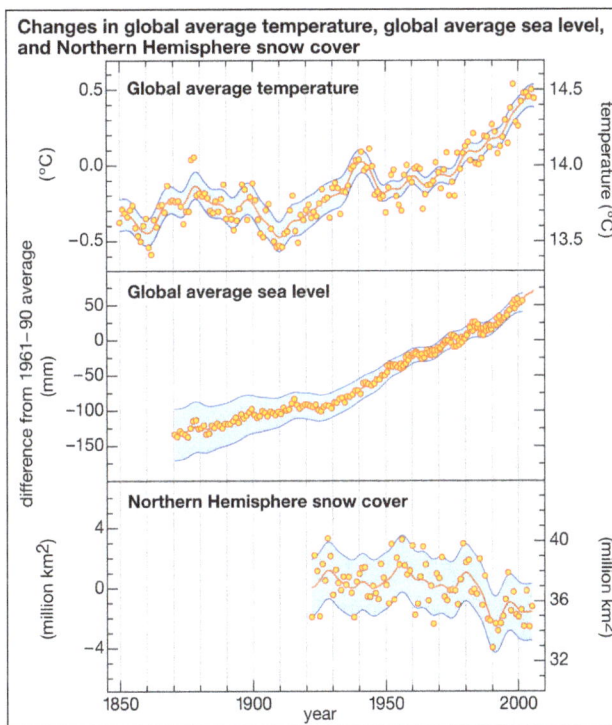

Changes in global average temperature, global average sea level, and Northern Hemisphere snow cover

Giving voice to a growing conviction of most of the scientific community, the Intergovernmental Panel on Climate Change (IPCC) was formed in 1988 by the World Meteorological Organization (WMO) and the United Nations Environment Program (UNEP). In 2013 the IPCC reported that the interval between 1880 and 2012 saw an increase in global average surface temperature of approximately 0.9 °C (1.5 °F). The increase is closer to 1.1 °C (2.0 °F) when measured relative to the preindustrial mean temperature.

A special report produced by the IPCC in 2018 honed this estimate further, noting that human beings and human activities have been responsible for a worldwide average temperature increase of between 0.8 and 1.2 °C (1.4 and 2.2 °F) of global warming since preindustrial times, and most of the warming observed over the second half of the 20th century could be attributed to human activities. It predicted that the global mean surface temperature would increase between 3 and 4 °C (5.4 and 7.2 °F) by 2100 relative to the 1986–2005 average should carbon emissions continue at their current rate. The predicted rise in temperature was based on a range of possible scenarios that accounted for future greenhouse gas emissions and mitigation (severity reduction) measures and on uncertainties in the model projections. Some of the main uncertainties include the precise role of feedback processes and the impacts of industrial pollutants known as aerosols, which may offset some warming.

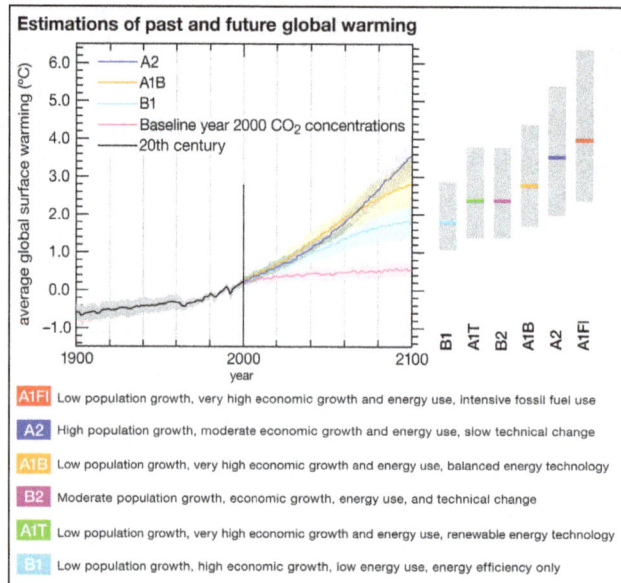

Estimations of past and future global warming

Global warming scenarios: Graph of the predicted increase in Earth's average surface temperature according to a series of climate change scenarios that assume different levels of economic development, population growth, and fossil fuel use. The assumptions made by each scenario are given at the bottom of the graph.

Many climate scientists agree that significant societal, economic, and ecological damage would result if global average temperatures rose by more than 2 °C (3.6 °F) in such a short time. Such damage would include increased extinction of many plant and animal species, shifts in patterns of agriculture, and rising sea levels. By 2015 all but a few national governments had begun the process of instituting carbon reduction plans as part of the Paris Agreement, a treaty designed to help countries keep global warming to 1.5 °C (2.7 °F) above preindustrial levels in order to avoid the worst of the predicted effects. Authors of a special report published by the IPCC in 2018 noted that should carbon emissions continue at their present rate, the increase in average near-surface air temperatures would reach 1.5 °C sometime between 2030 and 2052. Past IPCC assessments reported that the global average sea level rose by some 19–21 cm (7.5–8.3 inches) between 1901 and 2010 and that sea levels rose faster in the second half of the 20th century than in the first half. It also predicted, again depending on a wide range of scenarios, that the global average sea level would rise 26–77 cm (10.2–30.3 inches) relative to the 1986–2005 average by 2100 for global warming of 1.5 °C, an average of 10 cm (3.9 inches) less than what would be expected if warming rose to 2 °C (3.6 °F) above preindustrial levels.

The scenarios referred to above depend mainly on future concentrations of certain trace gases, called greenhouse gases, that have been injected into the lower atmosphere in increasing amounts through the burning of fossil fuels for industry, transportation, and residential uses. Modern global warming is the result of an increase in magnitude of the so-called greenhouse effect, a warming of Earth's surface and lower atmosphere caused by the presence of water vapour, carbon dioxide, methane, nitrous oxides, and other greenhouse gases.

Some incoming sunlight is reflected by Earth's atmosphere and surface, but most is absorbed by the surface, which is warmed. Infrared (IR) radiation is then emitted from the surface. Some IR radiation escapes to space, but some is absorbed by the atmosphere's greenhouse gases (especially water vapour, carbon dioxide, and methane) and reradiated in all directions, some to space and some back toward the surface, where it further warms the surface and the lower atmosphere.

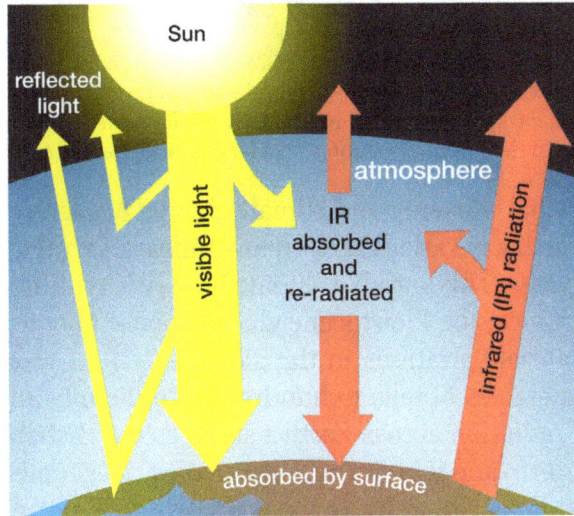

Greenhouse effect on Earth.

Of all these gases, carbon dioxide is the most important, both for its role in the greenhouse effect and for its role in the human economy. It has been estimated that, at the beginning of the industrial age in the mid-18th century, carbon dioxide concentrations in the atmosphere were roughly 280 parts per million (ppm). By the middle of 2018 they had risen to 406 ppm, and, if fossil fuels continue to be burned at current rates, they are projected to reach 550 ppm by the mid-21st century—essentially, a doubling of carbon dioxide concentrations in 300 years.

A vigorous debate is in progress over the extent and seriousness of rising surface temperatures, the effects of past and future warming on human life, and the need for action to reduce future warming and deal with its consequences. background and public policy debate related to the subject of global warming. It considers the causes of rising near-surface air temperatures, the influencing factors, the process of climate research and forecasting, the possible ecological and social impacts of rising temperatures, and the public policy developments since the mid-20th century.

Climatic Variation since the Last Glaciation

Global warming is related to the more general phenomenon of climate change, which refers to changes in the totality of attributes that define climate. In addition to changes in airtemperature, climate change involves changes to precipitation patterns, winds, ocean currents, and other measures of Earth's climate. Normally, climate change can be viewed as the combination of various natural forces occurring over diverse timescales. Since the advent of human civilization, climate change has involved an "anthropogenic," or exclusively human-caused, element, and this anthropogenic element has become more important in the industrial period of the past two centuries. The term *global warming* is used specifically to refer to any warming of near-surface air during the past two centuries that can be traced to anthropogenic causes.

A series of photographs of the Grinnell Glacier taken from the summit of Mount Gould in
Glacier National Park, Montana,. In 1938 the Grinnell Glacier filled the entire area
at the bottom of the image. By 2006 it had largely disappeared from this view.

To define the concepts of global warming and climate change properly, it is first necessary to rec-
ognize that the climate of Earth has varied across many timescales, ranging from an individual human life span to billions of years. This variable climate history is typically classified in terms of "regimes" or "epochs." For instance, the Pleistocene glacial epoch (about 2,600,000 to 11,700 years
ago) was marked by substantial variations in the global extent of glaciers and icesheets. These
variations took place on timescales of tens to hundreds of millennia and were driven by changes
in the distribution of solar radiation across Earth's surface. The distribution of solar radiation is
known as the insolation pattern, and it is strongly affected by the geometry of Earth's orbit around
the Sun and by the orientation, or tilt, of Earth's axis relative to the direct rays of the Sun.

Worldwide, the most recent glacial period, or ice age, culminated about 21,000 years ago in what is
often called the Last Glacial Maximum. During this time, continental ice sheets extended well into the
middle latitude regions of Europe and North America, reaching as far south as present-day London
and New York City. Global annual mean temperature appears to have been about 4–5 °C (7–9 °F)
colder than in the mid-20th century. It is important to remember that these figures are a global average. In fact, during the height of this last ice age, Earth's climate was characterized by greater cooling
at higher latitudes (that is, toward the poles) and relatively little cooling over large parts of the tropical
oceans (near the Equator). This glacial interval terminated abruptly about 11,700 years ago and was followed by the subsequent relatively ice-free period known as the Holocene Epoch. The modern period
of Earth's history is conventionally defined as residing within the Holocene. However, some scientists
have argued that the Holocene Epoch terminated in the relatively recent past and that Earth currently resides in a climatic interval that could justly be called the Anthropocene Epoch—that is, a period
during which humans have exerted a dominant influence over climate.

Though less dramatic than the climate changes that occurred during the Pleistocene Epoch, significant variations in global climate have nonetheless taken place over the course of the Holocene.
During the early Holocene, roughly 9,000 years ago, atmospheric circulation and precipitation
patterns appear to have been substantially different from those of today. For example, there is
evidence for relatively wet conditions in what is now the Sahara Desert. The change from one climatic regime to another was caused by only modest changes in the pattern of insolation within the
Holocene interval as well as the interaction of these patterns with large-scale climate phenomena
such as monsoons and El Niño/Southern Oscillation (ENSO).

During the middle Holocene, some 5,000–7,000 years ago, conditions appear to have been relatively warm—indeed, perhaps warmer than today in some parts of the world and during certain seasons.
For this reason, this interval is sometimes referred to as the Mid-Holocene Climatic Optimum. The

relative warmth of average near-surface air temperatures at this time, however, is somewhat unclear. Changes in the pattern of insolation favoured warmer summers at higher latitudes in the Northern Hemisphere, but these changes also produced cooler winters in the Northern Hemisphere and relatively cool conditions year-round in the tropics. Any overall hemispheric or global mean temperature changes thus reflected a balance between competing seasonal and regional changes. In fact, recent theoretical climate model studies suggest that global mean temperatures during the middle Holocene were probably 0.2–0.3 °C (0.4–0.5 °F) colder than average late 20th-century conditions.

Over subsequent millennia, conditions appear to have cooled relative to middle Holocene levels. This period has sometimes been referred to as the "Neoglacial." In the middle latitudes this cooling trend was associated with intermittent periods of advancing and retreating mountain glaciers reminiscent of (though far more modest than) the more substantial advance and retreat of the major continental ice sheets of the Pleistocene climate epoch.

Radiative Forcing

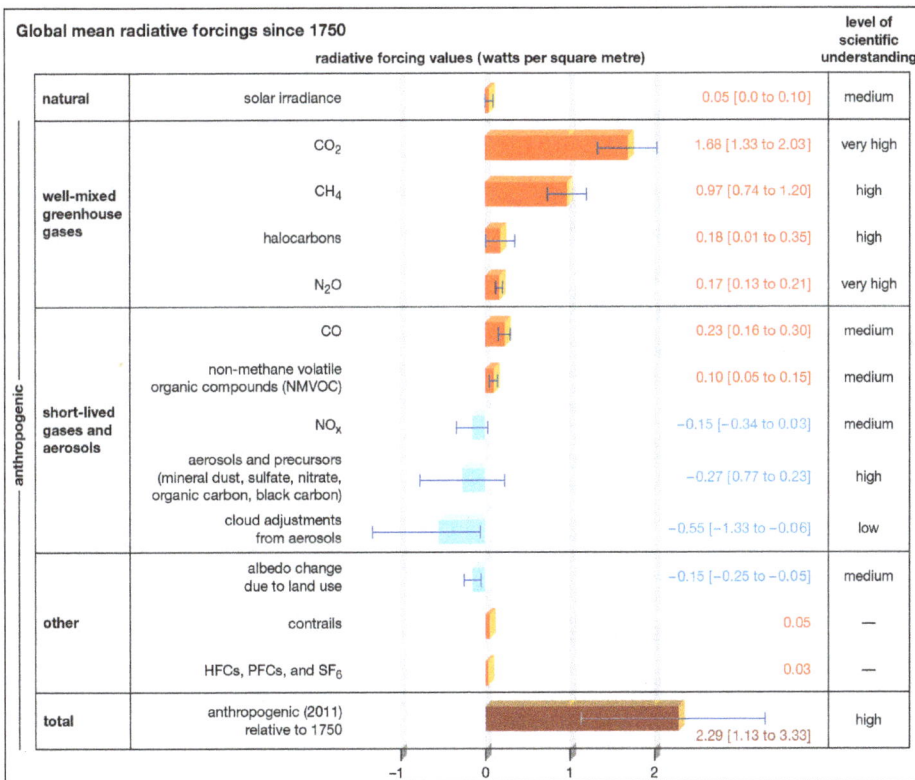

Global mean radiative forcings since 1750		radiative forcing values (watts per square metre)	level of scientific understanding
natural	solar irradiance	0.05 [0.0 to 0.10]	medium
well-mixed greenhouse gases	CO_2	1.66 [1.33 to 2.03]	very high
	CH_4	0.97 [0.74 to 1.20]	high
	halocarbons	0.18 [0.01 to 0.35]	high
	N_2O	0.17 [0.13 to 0.21]	very high
short-lived gases and aerosols	CO	0.23 [0.16 to 0.30]	medium
	non-methane volatile organic compounds (NMVOC)	0.10 [0.05 to 0.15]	medium
	NO_x	−0.15 [−0.34 to 0.03]	medium
	aerosols and precursors (mineral dust, sulfate, nitrate, organic carbon, black carbon)	−0.27 [0.77 to 0.23]	high
	cloud adjustments from aerosols	−0.55 [−1.33 to −0.06]	low
other	albedo change due to land use	−0.15 [−0.25 to −0.05]	medium
	contrails	0.05	—
	HFCs, PFCs, and SF_6	0.03	—
total	anthropogenic (2011) relative to 1750	2.29 [1.13 to 3.33]	high

Since 1750 the concentration of carbon dioxide and other greenhouse gases has increased in Earth's atmosphere. As a result of these and other factors, Earth's atmosphere retains more heat than in the past.

In light of the greenhouse effect, it is apparent that the temperature of Earth's surface and lower atmosphere may be modified in three ways: (1) through a net increase in the solar radiation entering at the top of Earth's atmosphere, (2) through a change in the fraction of the radiation reaching the surface, and (3) through a change in the concentration of greenhouse gases in the atmosphere. In each case the changes can be thought of in terms of "radiative forcing." As defined by the IPCC, radiative forcing is a measure of the influence a given climatic factor has on the amount of downward-directed radiant energy impinging upon Earth's surface. Climatic factors are divided between those caused primarily by human activity (such as greenhouse gas emissions and aerosol emissions) and those caused by natural forces (such as solar irradiance); then, for each factor, so-called forcing values are calculated for the time period between 1750 and the present day. "Positive forcing" is exerted by climatic factors that contribute to the warming of Earth's surface, whereas "negative forcing" is exerted by factors that cool Earth's surface.

On average, about 342 watts of solar radiation strike each square metre of Earth's surface per year, and this quantity can in turn be related to a rise or fall in Earth's surface temperature. Temperatures at the surface may also rise or fall through a change in the distribution of terrestrial radiation (that is, radiation emitted by Earth) within the atmosphere. In some cases, radiative forcing has a natural origin, such as during explosive eruptions from volcanoes where vented gases and ash block some portion of solar radiation from the surface. In other cases, radiative forcing has an anthropogenic, or exclusively human, origin. For example, anthropogenic increases in carbon dioxide, methane, and nitrous oxide are estimated to account for 2.3 watts per square metre of positive radiative forcing. When all values of positive and negative radiative forcing are taken together and all interactions between climatic factors are accounted for, the total net increase in surface radiation due to human activities since the beginning of the Industrial Revolution is 1.6 watts per square metre.

The Influences of Human Activity on Climate

Human activity has influenced global surface temperatures by changing the radiative balance governing the Earth on various timescales and at varying spatial scales. The most profound and well-known anthropogenic influence is the elevation of concentrations of greenhouse gases in the atmosphere. Humans also influence climate by changing the concentrations of aerosols and ozone and by modifying the land cover of Earth's surface.

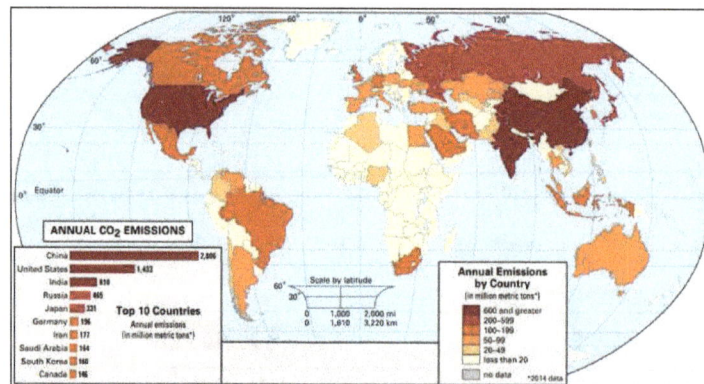

Carbon dioxide emissions: Map of annual carbon dioxide emissions by country in 2014.

Greenhouse Gases

Greenhouse gases warm Earth's surface by increasing the net downward longwave radiation reaching the surface. The relationship between atmospheric concentration of greenhouse gases and the associated positive radiative forcing of the surface is different for each gas. A complicated relationship exists between the chemical properties of each greenhouse gas and the relative amount of longwave radiation that each can absorb. What follows is a discussion of the radiative behaviour of each major greenhouse gas.

Factories that burn fossil fuels help to cause global warming.

Water Vapour

Water vapour is the most potent of the greenhouse gases in Earth's atmosphere, but its behaviour is fundamentally different from that of the other greenhouse gases. The primary role of water vapour is not as a direct agent of radiative forcing but rather as a climate feedback—that is, as a response within the climate system that influences the system's continued activity. This distinction arises from the fact that the amount of water vapour in the atmosphere cannot, in general, be directly modified by human behaviour but is instead set by air temperatures. The warmer the surface, the greater the evaporation rate of water from the surface. As a result, increased evaporation leads to a greater concentration of water vapour in the lower atmosphere capable of absorbing longwave radiation and emitting it downward.

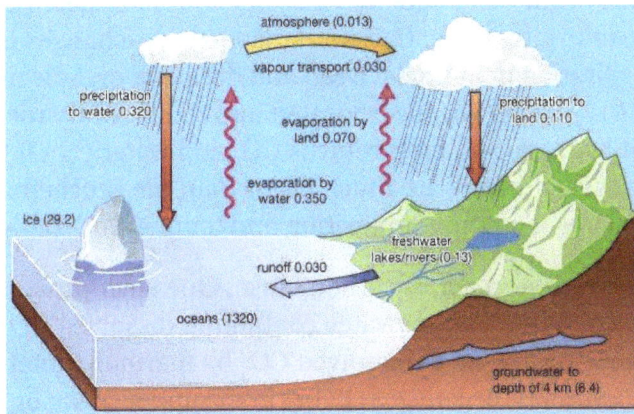

In the above figure the present-day surface hydrologic cycle, in which water is transferred from the oceans through the atmosphere to the continents and back to the oceans over and beneath the land surface. The values in parentheses following the various forms of water (e.g., ice) refer to volumes in millions of cubic kilometres; those following the processes (e.g., precipitation) refer to their fluxes in millions of cubic kilometres of water per year.

Carbon Dioxide

Of the greenhouse gases, carbon dioxide (CO_2) is the most significant. Natural sources of atmospheric CO_2 include outgassing from volcanoes, the combustion and natural decay of organic matter, and respiration by aerobic (oxygen-using) organisms. These sources are balanced, on average, by a set of physical, chemical, or biological processes, called "sinks," that tend to remove CO_2 from the atmosphere. Significant natural sinks include terrestrial vegetation, which takes up CO_2 during the process of photosynthesis.

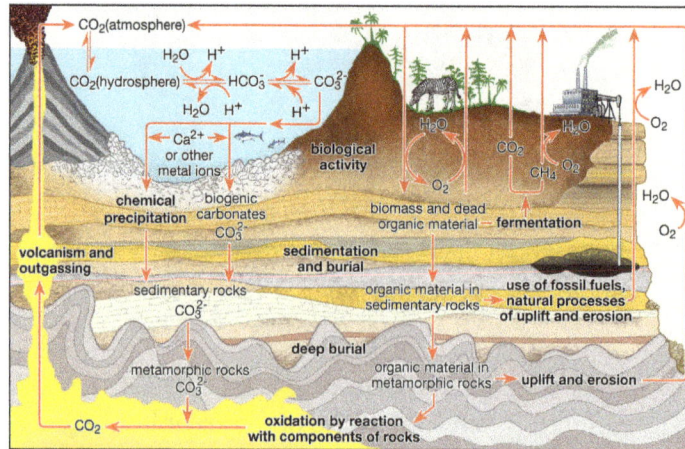

Carbon cycle.

Carbon is transported in various forms through the atmosphere, the hydrosphere, and geologic formations. One of the primary pathways for the exchange of carbon dioxide (CO_2) takes place between the atmosphere and the oceans; there a fraction of the CO_2 combines with water, forming carbonic acid (H_2CO_3) that subsequently loses hydrogen ions (H^+) to form bicarbonate (HCO_3^-) and carbonate (CO_3^{2-}) ions. Mollusk shells or mineral precipitates that form by the reaction of calcium or other metal ions with carbonate may become buried in geologic strata and eventually release CO_2 through volcanic outgassing. Carbon dioxide also exchanges through photosynthesis in plants and through respiration in animals. Dead and decaying organic matter may ferment and release CO_2 or methane (CH_4) or may be incorporated into sedimentary rock, where it is converted to fossil fuels. Burning of hydrocarbon fuels returns CO_2 and water (H_2O) to the atmosphere. The biological and anthropogenic pathways are much faster than the geochemical pathways and, consequently, have a greater impact on the composition and temperature of the atmosphere.

A number of oceanic processes also act as carbon sinks. One such process, called the "solubility pump," involves the descent of surface seawater containing dissolved CO_2. Another process, the "biological pump," involves the uptake of dissolved CO_2 by marine vegetation and phytoplankton (small free-floating photosynthetic organisms) living in the upper ocean or by other marine organisms that use CO_2 to build skeletons and other structures made of calcium carbonate ($CaCO_3$). As these organisms expire and fall to the ocean floor, the carbon they contain is transported downward and eventually buried at depth. A long-term balance between these natural sources and sinks leads to the background, or natural, level of CO_2 in the atmosphere.

In contrast, human activities increase atmospheric CO_2 levels primarily through the burning of fossil fuels—principally oil and coal and secondarily natural gas, for use in transportation, heating, and the generation of electrical power—and through the production of cement. Other anthropogenic sources include the burning of forests and the clearing of land. Anthropogenic emissions currently account for the annual release of about 7 gigatons (7 billion tons) of carbon into the atmosphere. Anthropogenic emissions are equal to approximately 3 percent of the total emissions of CO_2 by natural sources, and this amplified carbon load from human activities far exceeds the offsetting capacity of natural sinks (by perhaps as much as 2–3 gigatons per year).

Deforestation: Smoldering remains of a plot of deforested land in the Amazon Rainforest of Brazil. Annually, it is estimated that net global deforestation accounts for about two gigatons of carbon emissions to the atmosphere.

CO_2 consequently accumulated in the atmosphere at an average rate of 1.4 ppm per year between 1959 and 2006 and roughly 2.0 ppm per year between 2006 and 2018. Overall, this rate of accumulation has been linear (that is, uniform over time). However, certain current sinks, such as the oceans, could become sources in the future. This may lead to a situation in which the concentration of atmospheric CO_2 builds at an exponential rate (that is, its rate of increase is also increasing).

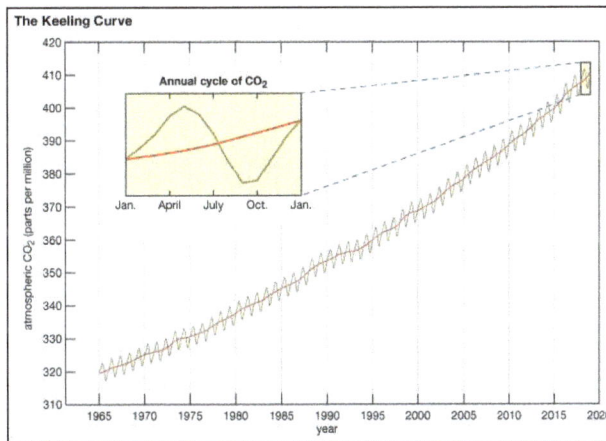

Figure show the Keeling Curve, named after American climate scientist Charles David Keeling, tracks changes in the concentration of carbon dioxide (CO_2) in Earth's atmosphere at a research station on Mauna Loa in Hawaii. Although these concentrations experience small seasonal fluctuations, the overall trend shows that CO_2 is increasing in the atmosphere.

The natural background level of carbon dioxide varies on timescales of millions of years because of slow changes in outgassing through volcanic activity. For example, roughly 100 million years ago, during the Cretaceous Period (145 million to 66 million years ago), CO_2 concentrations appear to have been several times higher than they are today (perhaps close to 2,000 ppm). Over the past 700,000 years, CO_2 concentrations have varied over a far smaller range (between roughly 180 and 300 ppm) in association with the same Earth orbital effects linked to the coming and going of the Pleistocene ice ages. By the early 21st century, CO_2 levels had reached 384 ppm, which is approximately 37 percent above the natural background level of roughly 280 ppm that existed at the beginning of the Industrial Revolution. Atmospheric CO_2 levels continued to increase, and by 2018 they had reached 410 ppm. Such levels are believed to be the highest in at least 800,000 years according to ice core measurements and may be the highest in at least 5 million years according to other lines of evidence.

Radiative forcing caused by carbon dioxide varies in an approximately logarithmic fashion with the concentration of that gas in the atmosphere. The logarithmic relationship occurs as the result of a saturation effect wherein it becomes increasingly difficult, as CO_2 concentrations increase, for additional CO_2 molecules to further influence the "infrared window" (a certain narrow band of wavelengths in the infrared region that is not absorbed by atmospheric gases). The logarithmic relationship predicts that the surface warming potential will rise by roughly the same amount for each doubling of CO_2 concentration. At current rates of fossil fuel use, a doubling of CO_2 concentrations over preindustrial levels is expected to take place by the middle of the 21st century (when CO_2 concentrations are projected to reach 560 ppm). A doubling of CO_2 concentrations would represent an increase of roughly 4 watts per square metre of radiative forcing. Given typical estimates of "climate sensitivity" in the absence of any offsetting factors, this energy increase would lead to a warming of 2 to 5 °C (3.6 to 9 °F) over preindustrial times. The total radiative forcing by anthropogenic CO_2 emissions since the beginning of the industrial age is approximately 1.66 watts per square metre.

Methane

Methane (CH_4) is the second most important greenhouse gas. CH_4 is more potent than CO_2 because the radiative forcing produced per molecule is greater. In addition, the infrared window is less saturated in the range of wavelengths of radiation absorbed by CH_4, so more molecules may fill in the region. However, CH_4 exists in far lower concentrations than CO_2 in the atmosphere, and its concentrations by volume in the atmosphere are generally measured in parts per billion (ppb) rather than ppm. CH_4 also has a considerably shorter residence time in the atmosphere than CO_2 (the residence time for CH_4 is roughly 10 years, compared with hundreds of years for CO_2).

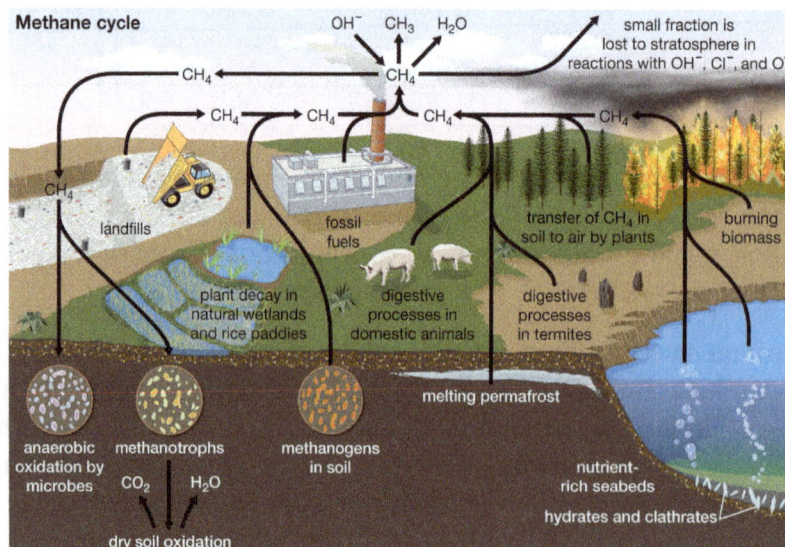

Methane cycle.

Natural sources of methane include tropical and northern wetlands, methane-oxidizing bacteria that feed on organic material consumed by termites, volcanoes, seepage vents of the seafloor in regions rich with organic sediment, and methane hydrates trapped along the continental shelves of the oceans and in polar permafrost. The primary natural sink for methane is the atmosphere itself, as methane reacts readily with the hydroxyl radical (OH) within the troposphere to form CO_2 and water vapour (H_2O). When CH_4 reaches the stratosphere, it is destroyed. Another natural sink is soil, where methane is oxidized by bacteria.

As with CO_2, human activity is increasing the CH_4 concentration faster than it can be offset by natural sinks. Anthropogenic sources currently account for approximately 70 percent of total annual emissions, leading to substantial increases in concentration over time. The major anthropogenic sources of atmospheric CH_4 are rice cultivation, livestock farming, the burning of coal and natural gas, the combustion of biomass, and the decomposition of organic matter in landfills. Future trends are particularly difficult to anticipate. This is in part due to an incomplete understanding of the climate feedbacks associated with CH_4 emissions. In addition it is difficult to predict how, as human populations grow, possible changes in livestock raising, rice cultivation, and energy utilization will influence CH_4 emissions.

It is believed that a sudden increase in the concentration of methane in the atmosphere was responsible for a warming event that raised average global temperatures by 4–8 °C (7.2–14.4 °F) over a few thousand years during the so-called Paleocene-Eocene Thermal Maximum, or PETM. This episode took place roughly 55 million years ago, and the rise in CH_4 appears to have been related to a massive volcanic eruption that interacted with methane-containing flood deposits. As a result, large amounts of gaseous CH_4 were injected into the atmosphere. It is difficult to know precisely how high these concentrations were or how long they persisted. At very high concentrations, residence times of CH_4 in the atmosphere can become much greater than the nominal 10-year residence time that applies today. Nevertheless, it is likely that these concentrations reached several ppm during the PETM.

Methane concentrations have also varied over a smaller range (between roughly 350 and 800 ppb) in association with the Pleistocene ice age cycles. Preindustrial levels of CH_4 in the atmosphere were approximately 700 ppb, whereas levels exceeded 1,867 ppb in late 2018. (These concentrations are well above the natural levels observed for at least the past 650,000 years.) The net radiative forcing by anthropogenic CH_4 emissions is approximately 0.5 watt per square metre—or roughly one-third the radiative forcing of CO_2.

Surface-level Ozone and other Compounds

The next most significant greenhouse gas is surface, or low-level, ozone (O_3). Surface O_3 is a result of air pollution; it must be distinguished from naturally occurring stratospheric O_3, which has a very different role in the planetary radiation balance. The primary natural source of surface O_3 is the subsidence of stratospheric O_3 from the upper atmosphere. In contrast, the primary anthropogenic source of surface O_3 is photochemical reactions involving the atmospheric pollutant carbon monoxide (CO). The best estimates of the natural concentration of surface O_3 are 10 ppb, and the net radiative forcing due to anthropogenic emissions of surface O_3 is approximately 0.35 watt per square metre. Ozone concentrations can rise above unhealthy levels (that is, conditions where concentrations meet or exceed 70 ppb for eight hours or longer) in cities prone to photochemical smog.

Nitrous Oxides and Fluorinated Gases

Additional trace gases produced by industrial activity that have greenhouse properties include nitrous oxide (N_2O) and fluorinated gases (halocarbons), the latter including sulfur hexafluoride, hydrofluorocarbons (HFCs), and perfluorocarbons (PFCs). Nitrous oxide is responsible for 0.16 watt per square metre radiative forcing, while fluorinated gases are collectively responsible for 0.34 watt per square metre. Nitrous oxides have small background concentrations due to natural biological reactions in soil and water, whereas the fluorinated gases owe their existence almost entirely to industrial sources.

Aerosols

The production of aerosols represents an important anthropogenic radiative forcing of climate. Collectively, aerosols block—that is, reflect and absorb—a portion of incoming solar radiation, and this creates a negative radiative forcing. Aerosols are second only to greenhouse gases in relative importance in their impact on near-surface air temperatures. Unlike the decade-long residence times of the "well-mixed" greenhouse gases, such as CO_2 and CH_4, aerosols are readily flushed out of the atmosphere within days, either by rain or snow (wet deposition) or by settling out of the air (dry deposition). They must therefore be continually generated in order to produce a steady effect on radiative forcing. Aerosols have the ability to influence climate directly by absorbing or reflecting incoming solar radiation, but they can also produce indirect effects on climate by modifying cloud formation or cloud properties. Most aerosols serve as condensation nuclei (surfaces upon which water vapour can condense to form clouds); however, darker-coloured aerosols may hinder cloud formation by absorbing sunlight and heating up the surrounding air. Aerosols can be transported thousands of kilometres from their sources of origin by winds and upper-level circulation in the atmosphere.

Perhaps the most important type of anthropogenic aerosol in radiative forcing is sulfateaerosol. It is produced from sulfur dioxide (SO_2) emissions associated with the burning of coal and oil. Since the late 1980s, global emissions of SO_2 have decreased from about 151.5 million tonnes (167.0 million tons) to less than 100 million tonnes (110.2 million tons) of sulfur per year.

Nitrate aerosol is not as important as sulfate aerosol, but it has the potential to become a significant source of negative forcing. One major source of nitrate aerosol is smog (the combination of ozone with oxides of nitrogen in the lower atmosphere) released from the incomplete burning of fuel in internal-combustion engines. Another source is ammonia (NH_3), which is often used in fertilizers or released by the burning of plants and other organic materials. If greater amounts of atmospheric nitrogen are converted to ammonia and agricultural ammonia emissions continue to increase as projected, the influence of nitrate aerosols on radiative forcing is expected to grow.

Both sulfate and nitrate aerosols act primarily by reflecting incoming solar radiation, thereby reducing the amount of sunlight reaching the surface. Most aerosols, unlike greenhouse gases, impart a cooling rather than warming influence on Earth's surface. One prominent exception is carbonaceous aerosols such as carbon black or soot, which are produced by the burning of fossil fuels and biomass. Carbon black tends to absorb rather than reflect incident solar radiation, and so it has a warming impact on the lower atmosphere, where it resides. Because of its absorptive properties, carbon black is also capable of having an additional indirect effect on climate. Through its deposition in snowfall, it can decrease the albedo of snow cover. This reduction in the amount of solar radiation reflected back to space by snow surfaces creates a minor positive radiative forcing.

Natural forms of aerosol include windblown mineral dust generated in arid and semiarid regions and sea salt produced by the action of waves breaking in the ocean. Changes to windpatterns as a result of climate modification could alter the emissions of these aerosols. The influence of climate change on regional patterns of aridity could shift both the sources and the destinations of dust clouds. In addition, since the concentration of sea salt aerosol, or sea aerosol, increases with the strength of the winds near the ocean surface, changes in wind speed due to global warming and climate change could influence the concentration of sea salt aerosol. For example, some studies suggest that climate change might lead to stronger winds over parts of the North Atlantic Ocean. Areas with stronger winds may experience an increase in the concentration of sea salt aerosol.

Other natural sources of aerosols include volcanic eruptions, which produce sulfate aerosol, and biogenic sources (e.g., phytoplankton), which produce dimethyl sulfide (DMS). Other important biogenic aerosols, such as terpenes, are produced naturally by certain kinds of treesor other plants. For example, the dense forests of the Blue Ridge Mountains of Virginia in the United States emit terpenes during the summer months, which in turn interact with the high humidity and warm temperatures to produce a natural photochemical smog. Anthropogenic pollutants such as nitrate and ozone, both of which serve as precursor molecules for the generation of biogenic aerosol, appear to have increased the rate of production of these aerosols severalfold. This process appears to be responsible for some of the increased aerosol pollution in regions undergoing rapid urbanization.

Human activity has greatly increased the amount of aerosol in the atmosphere compared with the background levels of preindustrial times. In contrast to the global effects of greenhouse gases, the impact of anthropogenic aerosols is confined primarily to the Northern Hemisphere, where most of the world's industrial activity occurs. The pattern of increases in anthropogenic aerosol over time is also somewhat different from that of greenhouse gases. During the middle of the 20th century, there was a substantial increase in aerosol emissions. This appears to have been at least partially responsible for a cessation of surface warming that took place in the Northern Hemisphere from the 1940s through the 1970s. Since that time, aerosol emissions have leveled off due to antipollution measures undertaken in the industrialized countries since the 1960s. Aerosol emissions may rise in the future, however, as a result of the rapid emergence of coal-fired electric power generation in China and India.

The total radiative forcing of all anthropogenic aerosols is approximately −1.2 watts per square metre. Of this total, −0.5 watt per square metre comes from direct effects (such as the reflection of solar energy back into space), and −0.7 watt per square metre comes from indirect effects (such as the influence of aerosols on cloud formation). This negative radiative forcing represents an offset of roughly 40 percent from the positive radiative forcing caused by human activity. However, the relative uncertainty in aerosol radiative forcing (approximately 90 percent) is much greater than that of greenhouse gases. In addition, future emissions of aerosols from human activities, and the influence of these emissions on future climate change, are not known with any certainty. Nevertheless, it can be said that, if concentrations of anthropogenic aerosols continue to decrease as they have since the 1970s, a significant offset to the effects of greenhouse gases will be reduced, opening future climate to further warming.

Land-use Change

There are a number of ways in which changes in land use can influence climate. The most direct influence is through the alteration of Earth's albedo, or surface reflectance. For example, the replacement of forest by cropland and pasture in the middle latitudes over the past several centuries has led to an increase in albedo, which in turn has led to greater reflection of incoming solar radiation in those regions. This replacement of forest by agriculture has been associated with a change in global average radiative forcing of approximately −0.2 watt per square metre since 1750. In Europe and other major agricultural regions, such land-use conversion began more than 1,000 years ago and has proceeded nearly to completion. For Europe, the negative radiative forcing due to land-use change has probably been substantial, perhaps approaching −5 watts per square metre. The influence of early land use on radiative forcing may help to explain a long period of cooling in Europe that followed a period of relatively mild conditions roughly 1,000 years ago. It is generally believed that the mild temperatures of this "medieval warm period," which was followed by a long period of cooling, rivaled those of 20th-century Europe.

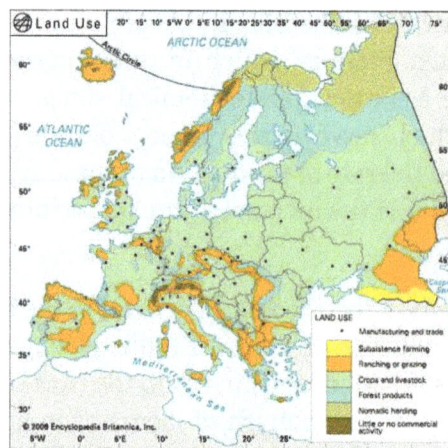

Europe: Land use Land use.

Land-use changes can also influence climate through their influence on the exchange of heat between Earth's surface and the atmosphere. For example, vegetation helps to facilitate the evaporation of water into the atmosphere through evapotranspiration. In this process, plants take up liquid water from the soil through their root systems. Eventually this water is released through transpiration into the atmosphere, as water vapour through the stomata in leaves. While deforestation generally leads to surface cooling due to the albedo factor, the land surface may also be warmed as a result of the release of latent heat by the evapotranspiration process. The relative importance of these two factors, one exerting a cooling effect and the other a warming effect, varies by both season and region. While the albedo effect is likely to dominate in middle latitudes, especially during the period from autumn through spring, the evapotranspiration effect may dominate during the summer in the midlatitudes and year-round in the tropics. The latter case is particularly important in assessing the potential impacts of continued tropical deforestation.

The rate at which tropical regions are deforested is also relevant to the process of carbon sequestration, the long-term storage of carbon in underground cavities and biomass rather than in the atmosphere. By removing carbon from the atmosphere, carbon sequestration acts to mitigate global warming. Deforestation contributes to global warming, as fewer plants are available to take up carbon dioxide from the atmosphere. In addition, as fallen trees, shrubs, and other plants are burned or allowed to slowly decompose, they release as carbon dioxide the carbon they stored during their lifetimes. Furthermore, any land-use change that influences the amount, distribution, or type of vegetation in a region can affect the concentrations of biogenic aerosols, though the impact of such changes on climate is indirect and relatively minor.

Stratospheric Ozone Depletion

Since the 1970s the loss of ozone (O_3) from the stratosphere has led to a small amount of negative radiative forcing of the surface. This negative forcing represents a competition between two distinct effects caused by the fact that ozone absorbs solar radiation. In the first case, as ozone levels in the stratosphere are depleted, more solar radiation reaches Earth's surface. In the absence of any other influence, this rise in insolation would represent a positive radiative forcing of the surface. However, there is a second effect of ozone depletion that is related to its greenhouse properties. As the amount of ozone in the stratosphere is decreased, there is also less ozone to absorb longwave radiation emitted by Earth's surface. With less absorption of radiation by ozone, there is a corresponding decrease in the downward reemission of radiation. This second effect overwhelms the first and results in a modest negative radiative forcing of Earth's surface and a modest cooling of the lower stratosphere by approximately 0.5 °C (0.9 °F) per decade since the 1970s.

Natural Influences on Climate

There are a number of natural factors that influence Earth's climate. These factors include external influences such as explosive volcanic eruptions, natural variations in the output of the Sun, and slow changes in the configuration of Earth's orbit relative to the Sun. In addition, there are natural oscillations in Earth's climate that alter global patterns of wind circulation, precipitation, and surface temperatures. One such phenomenon is the El Niño/Southern Oscillation (ENSO), a coupled atmospheric and oceanic event that occurs in the Pacific Oceanevery three to seven years. In addition, the Atlantic Multidecadal Oscillation (AMO) is a similar phenomenon that occurs over decades in the North Atlantic Ocean. Other types of oscillatory behaviour that produce dramatic shifts in climate may occur across timescales of centuries and millennia.

Volcanic Aerosols

Explosive volcanic eruptions have the potential to inject substantial amounts of sulfate aerosols into the lower stratosphere. In contrast to aerosol emissions in the lower troposphere, aerosols that enter the stratosphere may remain for several years before settling out, because of the relative absence of turbulent motions there. Consequently, aerosols from explosive volcanic eruptions have the potential to affect Earth's climate. Less-explosive eruptions, or eruptions that are less vertical in orientation, have a lower potential for substantial climate impact. Furthermore, because of large-scale circulation patterns within the stratosphere, aerosols injected within tropical regions tend to spread out over the globe, whereas aerosols injected within midlatitude and polar regions tend to remain confined to the middle and high latitudes of that hemisphere. Tropical eruptions, therefore, tend to have a greater climatic impact than eruptions occurring toward the poles. In 1991 the moderate eruption of Mount Pinatubo in the Philippines provided a peak forcing of approximately −4 watts per square metre and cooled the climate by about 0.5 °C (0.9 °F) over the following few years.

By comparison, the 1815 Mount Tambora eruption in present-day Indonesia, typically implicated for the 1816 "year without a summer" in Europe and North America, is believed to have been associated with a radiative forcing of approximately −6 watts per square metre.

A column of gas and ash rising from Mount Pinatubo in the Philippines on just days before the volcano's climactic explosion.

While in the stratosphere, volcanic sulfate aerosol actually absorbs longwave radiation emitted by Earth's surface, and absorption in the stratosphere tends to result in a cooling of the troposphere below. This vertical pattern of temperature change in the atmosphere influences the behaviour of winds in the lower atmosphere, primarily in winter. Thus, while there is essentially a global cooling effect for the first few years following an explosive volcanic eruption, changes in the winter patterns of surface winds may actually lead to warmer winters in some areas, such as Europe. Some modern examples of major eruptions include Krakatoa(Indonesia) in 1883, El Chichón (Mexico) in 1982, and Mount Pinatubo in 1991. There is also evidence that volcanic eruptions may influence other climate phenomena such as ENSO.

Variations in Solar Output

Direct measurements of solar irradiance, or solar output, have been available from satellites only since the late 1970s. These measurements show a very small peak-to-peak variation in solar irradiance (roughly 0.1 percent of the 1,366 watts per square metre received at the top of the atmosphere, for approximately 1.4 watts per square metre). However, indirect measures of solar activity are available from historical sunspot measurements dating back through the early 17th century. Attempts have been made to reconstruct graphs of solar irradiance variations from historical sunspot data by calibrating them against the measurements from modern satellites. However, since the modern measurements span only a few of the most recent 11-year solar cycles, estimates of solar output variability on 100-year and longer timescales are poorly correlated. Different assumptions regarding the relationship between the amplitudes of 11-year solar cycles and long-period solar output changes can lead to considerable differences in the resulting solar reconstructions. These differences in turn lead to fairly large uncertainty in estimating positive forcing by changes in solar irradiance since 1750. (Estimates range from 0.06 to 0.3 watt per square metre.) Even more challenging, given the lack of any modern analog, is the estimation of solar irradiance during the so-called Maunder Minimum, a period lasting from the mid-17th century to the early 18th century when very few sunspots were observed. While it is likely that solar irradiance was reduced at this time, it is difficult to calculate by how much. However, additional proxies of solar output exist that match reasonably well with the sunspot-derived records following the Maunder Minimum; these may be used as crude estimates of the solar irradiance variations.

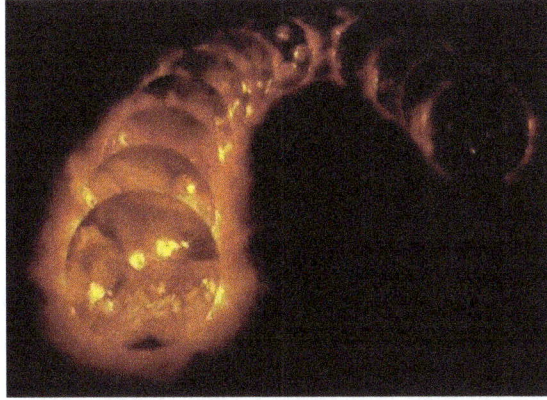

Twelve solar X-ray images obtained by Yohkoh between 1991 and 1995. The solar coronal brightness decreases by a factor of about 100 during a solar cycle as the Sun goes from an "active" state (left) to a less active state (right).

In theory it is possible to estimate solar irradiance even farther back in time, over at least the past millennium, by measuring levels of cosmogenic isotopes such as carbon-14 and beryllium-10. Cosmogenic isotopes are isotopes that are formed by interactions of cosmic rays with atomic nuclei in the atmosphere and that subsequently fall to Earth, where they can be measured in the annual layers found in ice cores. Since their production rate in the upper atmosphere is modulated by changes in solar activity, cosmogenic isotopes may be used as indirect indicators of solar irradiance. However, as with the sunspot data, there is still considerable uncertainty in the amplitude of past solar variability implied by these data.

Solar forcing also affects the photochemical reactions that manufacture ozone in the stratosphere. Through this modulation of stratospheric ozone concentrations, changes in solar irradiance (particularly in the ultraviolet portion of the electromagnetic spectrum) can modify how both shortwave and longwave radiation in the lower stratosphere are absorbed. As a result, the vertical temperature profile of the atmosphere can change, and this change can in turn influence phenomena such as the strength of the winter jet streams.

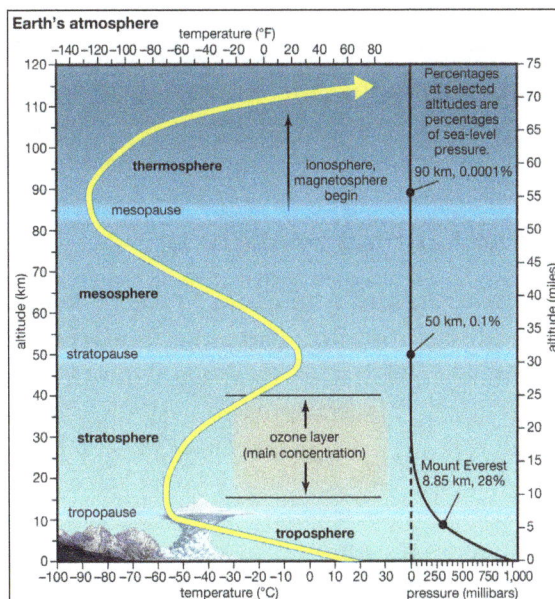

The layers of Earth's atmosphere. The yellow line shows the response of air temperature to increasing height.

Variations in Earth's Orbit

On timescales of tens of millennia, the dominant radiative forcing of Earth's climate is associated with slow variations in the geometry of Earth's orbit about the Sun. These variations include the precession of the equinoxes (that is, changes in the timing of summer and winter), occurring on a roughly 26,000-year timescale; changes in the tilt angle of Earth's rotational axis relative to the plane of Earth's orbit around the Sun, occurring on a roughly 41,000-year timescale; and changes in the eccentricity (the departure from a perfect circle) of Earth's orbit around the Sun, occurring on a roughly 100,000-year timescale. Changes in eccentricity slightly influence the mean annual solar radiation at the top of Earth's atmosphere, but the primary influence of all the orbital variations listed above is on the seasonal and latitudinal distribution of incoming solar radiation over Earth's surface. The major ice ages of the Pleistocene Epochwere closely related to the influence of these variations on summer insolation at high northern latitudes. Orbital variations thus exerted a primary control on the extent of continental ice sheets. However, Earth's orbital changes are generally believed to have had little impact on climate over the past few millennia, and so they are not considered to be significant factors in present-day climate variability.

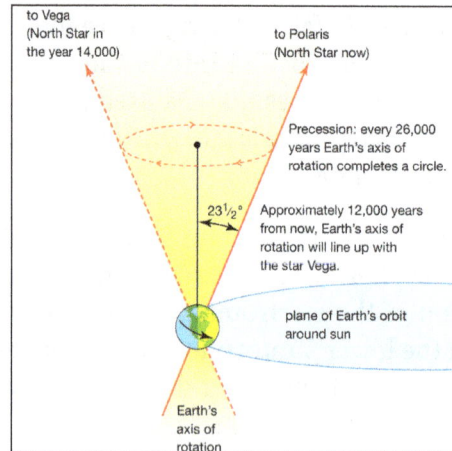

Earth's axis of rotation itself rotates, or precesses, completing one circle every 26,000 years. Consequently, Earth's North Pole points toward different stars (and sometimes toward empty space) as it travels in this circle. This precession is so slow that it is not noticeable in a person's lifetime, though astronomers must consider its effect when studying ancient sites such as Stonehenge.

Feedback Mechanisms and Climate Sensitivity

There are a number of feedback processes important to Earth's climate system and, in particular, its response to external radiative forcing. The most fundamental of these feedback mechanisms involves the loss of longwave radiation to space from the surface. Since this radiative loss increases with increasing surface temperatures according to the Stefan-Boltzmann law, it represents a stabilizing factor (that is, a negative feedback) with respect to near-surface air temperature.

Climate sensitivity can be defined as the amount of surface warming resulting from each additional watt per square metre of radiative forcing. Alternatively, it is sometimes defined as the warming that would result from a doubling of CO_2 concentrations and the associated addition of

4 watts per square metre of radiative forcing. In the absence of any additional feedbacks, climate sensitivity would be approximately 0.25 °C (0.45 °F) for each additional watt per square metre of radiative forcing. Stated alternatively, if the CO_2 concentration of the atmosphere present at the start of the industrial age (280 ppm) were doubled (to 560 ppm), the resulting additional 4 watts per square metre of radiative forcing would translate into a 1 °C (1.8 °F) increase in air temperature. However, there are additional feedbacks that exert a destabilizing, rather than stabilizing, influence, and these feedbacks tend to increase the sensitivity of climate to somewhere between 0.5 and 1.0 °C (0.9 and 1.8 °F) for each additional watt per square metre of radiative forcing.

Water Vapour Feedback

Unlike concentrations of other greenhouse gases, the concentration of water vapour in the atmosphere cannot freely vary. Instead, it is determined by the temperature of the lower atmosphere and surface through a physical relationship known as the Clausius-Clapeyron equation, named for 19th-century German physicist Rudolf Clausius and 19th-century French engineer Émile Clapeyron. Under the assumption that there is a liquid water surface in equilibrium with the atmosphere, this relationship indicates that an increase in the capacity of air to hold water vapour is a function of increasing temperature of that volume of air. This assumption is relatively good over the oceans, where water is plentiful, but not over the continents. For this reason the relative humidity (the percent of water vapour the air contains relative to its capacity) is approximately 100 percent over ocean regions and much lower over continental regions (approaching 0 percent in arid regions). Not surprisingly, the average relative humidity of Earth's lower atmosphere is similar to the fraction of Earth's surface covered by the oceans (that is, roughly 70 percent). This quantity is expected to remain approximately constant as Earth warms or cools. Slight changes to global relative humidity may result from human land-use modification, such as tropical deforestation and irrigation, which can affect the relative humidity over land areas up to regional scales.

The amount of water vapour in the atmosphere will rise as the temperature of the atmosphere rises. Since water vapour is a very potent greenhouse gas, even more potent than CO_2, the net greenhouse effect actually becomes stronger as the surface warms, which leads to even greater warming. This positive feedback is known as the "water vapour feedback." It is the primary reason that climate sensitivity is substantially greater than the previously stated theoretical value of 0.25 °C (0.45 °F) for each increase of 1 watt per square metre of radiative forcing.

Cloud Feedbacks

It is generally believed that as Earth's surface warms and the atmosphere's water vapour content increases, global cloud cover increases. However, the effects on near-surface air temperatures are complicated. In the case of low clouds, such as marine stratus clouds, the dominant radiative feature of the cloud is its albedo. Here any increase in low cloud cover acts in much the same way as an increase in surface ice cover: more incoming solar radiation is reflected and Earth's surface cools. On the other hand, high clouds, such as the towering cumulus clouds that extend up to the boundary between the troposphere and stratosphere, have a quite different impact on the surface radiation balance. The tops of cumulus clouds are considerably higher in the atmosphere and colder than their undersides. Cumulus cloud tops emit less longwave radiation out to space than the warmer cloud bottoms emit downward toward the surface. The end result of the formation of high cumulus clouds is greater warming at the surface.

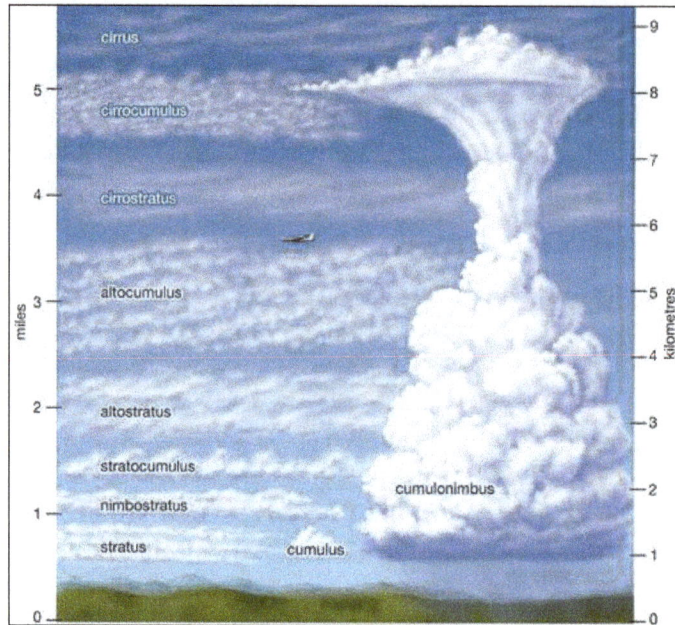

Different types of clouds form at different heights.

The net feedback of clouds on rising surface temperatures is therefore somewhat uncertain. It represents a competition between the impacts of high and low clouds, and the balance is difficult to determine. Nonetheless, most estimates indicate that clouds on the whole represent a positive feedback and thus additional warming.

Ice Albedo Feedback

Another important positive climate feedback is the so-called ice albedo feedback. This feedback arises from the simple fact that ice is more reflective (that is, has a higher albedo) than land or water surfaces. Therefore, as global ice cover decreases, the reflectivity of Earth's surface decreases, more incoming solar radiation is absorbed by the surface, and the surface warms. This feedback is considerably more important when there is relatively extensive global ice cover, such as during the height of the last ice age, roughly 25,000 years ago. On a global scale the importance of ice albedo feedback decreases as Earth's surface warms and there is relatively less ice available to be melted.

Carbon Cycle Feedbacks

Another important set of climate feedbacks involves the global carbon cycle. In particular, the two main reservoirs of carbon in the climate system are the oceans and the terrestrial biosphere. These reservoirs have historically taken up large amounts of anthropogenic CO_2 emissions. Roughly 50–70 percent is removed by the oceans, whereas the remainder is taken up by the terrestrial biosphere. Global warming, however, could decrease the capacity of these reservoirs to sequester atmospheric CO_2. Reductions in the rate of carbon uptake by these reservoirs would increase the pace of CO_2 buildup in the atmosphere and represent yet another possible positive feedback to increased greenhouse gas concentrations.

In the world's oceans, this feedback effect might take several paths. First, as surface waters warm, they would hold less dissolved CO_2. Second, if more CO_2 were added to the atmosphere and taken up by the oceans, bicarbonate ions (HCO_3^-) would multiply and ocean acidity would increase. Since calcium carbonate ($CaCO_3$) is broken down by acidic solutions, rising acidity would threaten ocean-dwelling fauna that incorporate $CaCO_3$ into their skeletons or shells. As it becomes increasingly difficult for these organisms to absorb oceanic carbon, there would be a corresponding decrease in the efficiency of the biological pump that helps to maintain the oceans as a carbon sink. Third, rising surface temperatures might lead to a slowdown in the so-called thermohaline circulation, a global pattern of oceanic flow that partly drives the sinking of surface waters near the poles and is responsible for much of the burial of carbon in the deep ocean. A slowdown in this flow due to an influx of melting fresh water into what are normally saltwater conditions might also cause the solubility pump, which transfers CO_2 from shallow to deeper waters, to become less efficient. Indeed, it is predicted that if global warming continued to a certain point, the oceans would cease to be a net sink of CO_2 and would become a net source.

As large sections of tropical forest are lost because of the warming and drying of regions such as Amazonia, the overall capacity of plants to sequester atmospheric CO_2 would be reduced. As a result, the terrestrial biosphere, though currently a carbon sink, would become a carbon source. Ambient temperature is a significant factor affecting the pace of photosynthesis in plants, and many plant species that are well adapted to their local climatic conditions have maximized their photosynthetic rates. As temperatures increase and conditions begin to exceed the optimal temperature range for both photosynthesis and soil respiration, the rate of photosynthesis would decline. As dead plants decompose, microbial metabolic activity (a CO_2 source) would increase and would eventually outpace photosynthesis.

Under sufficient global warming conditions, methane sinks in the oceans and terrestrial biosphere also might become methane sources. Annual emissions of methane by wetlands might either increase or decrease, depending on temperatures and input of nutrients, and it is possible that wetlands could switch from source to sink. There is also the potential for increased methane release as a result of the warming of Arctic permafrost (on land) and further methane release at the continental margins of the oceans (a few hundred metres below sea level). The current average atmospheric methane concentration of 1,750 ppb is equivalent to 3.5 gigatons (3.5 billion tons) of carbon. There are at least 400 gigatons of carbon equivalent stored in Arctic permafrost and as much as 10,000 gigatons (10 trillion tons) of carbon equivalent trapped on the continental margins of the oceans in a hydrated crystalline form known as clathrate. It is believed that some fraction of this trapped methane could become unstable with additional warming, although the amount and rate of potential emission remain highly uncertain.

Climate Research

Modern research into climatic variation and change is based on a variety of empirical and theoretical lines of inquiry. One line of inquiry is the analysis of data that record changes in atmosphere, oceans, and climate from roughly 1850 to the present. In a second line of inquiry, information describing paleoclimatic changes is gathered from "proxy," or indirect, sources such as ocean and lake sediments, pollen grains, corals, ice cores, and tree rings. Finally, a variety of theoretical models can be used to investigate the behaviour of Earth's climate under different conditions.

Modern Observations

Although a limited regional subset of land-based records is available from the 17th and 18th centuries, instrumental measurements of key climate variables have been collected systematically and at global scales since the mid-19th to early 20th century. These data include measurements of surface temperature on land and at sea, atmospheric pressure at sea level, precipitation over continents and oceans, sea ice extents, surface winds, humidity, and tides. Such records are the most reliable of all available climate data, since they are precisely dated and are based on well-understood instruments and physical principles. Corrections must be made for uncertainties in the data (for instance, gaps in the observational record, particularly during earlier years) and for systematic errors (such as an "urban heat island" bias in temperature measurements made on land).

Since the mid-20th century a variety of upper-air observations have become available (for example, of temperature, humidity, and winds), allowing climatic conditions to be characterized from the ground upward through the upper troposphere and lower stratosphere. Since the 1970s these data have been supplemented by polar-orbiting and geostationary satellites and by platforms in the oceans that gauge temperature, salinity, and other properties of seawater. Attempts have been made to fill the gaps in early measurements by using various statistical techniques and "backward prediction" models and by assimilatingavailable observations into numerical weather prediction models. These techniques seek to estimate meteorological observations or atmospheric variables (such as relative humidity) that have been poorly measured in the past.

Modern measurements of greenhouse gas concentrations began with an investigation of atmospheric carbon dioxide (CO_2) concentrations by American climate scientist Charles Keeling at the summit of Mauna Loa in Hawaii in 1958. Keeling's findings indicated that CO_2 concentrations were steadily rising in association with the combustion of fossil fuels, and they also yielded the famous "Keeling curve," a graph in which the longer-term rising trend is superimposed on small oscillations related to seasonal variations in the uptake and release of CO_2 from photosynthesis and respiration in the terrestrial biosphere. Keeling's measurements at Mauna Loa apply primarily to the Northern Hemisphere.

Taking into account the uncertainties, the instrumental climate record indicates substantial trends since the end of the 19th century consistent with a warming Earth. These trends include a rise in global surface temperature of 0.9 °C (1.5 °F) between 1880 and 2012, an associated elevation of global sea level of 19–21 cm (7.5–8.3 inches) between 1901 and 2010, and a decrease in snow cover in the Northern Hemisphere of approximately 1.5 million square km (580,000 square miles). Records of average global temperatures kept by the World Meteorological Organization (WMO) indicate that the years 1998, 2005, and 2010 are statistically tied with one another as the warmest years since modern record keeping began in 1880; the WMO also noted that the decade 2001–10 was the warmest decade since 1880. Increases in global sea level are attributed to a combination of seawater expansion due to ocean heating and freshwater runoff caused by the melting of terrestrial ice. Reductions in snow cover are the result of warmer temperatures favouring a steadily shrinking winter season.

Climate data collected during the first two decades of the 21st century reveal that surface warming between 2005 and 2014 proceeded slightly more slowly than was expected from the effect of greenhouse gas increases alone. This fact was sometimes used to suggest that global warming had stopped or that it experienced a "hiatus" or "pause." In reality, this phenomenon appears to have been influenced by several factors, none of which, however, implies that global warming stopped during this period or that global warming would not continue in the future. One factor was the increased burial of heat beneath the ocean surface by strong trade winds, a process assisted by La Niña conditions. The effects of La Niña manifest in the form of cooling surface waters along the western coast of South America. As a result, warming at the ocean surface was reduced, but the accumulation of heat in other parts of the ocean occurred at an accelerated rate. Another factor cited by climatologists was a small but potentially important increase in aerosols from volcanic activity, which may have blocked a small portion of incoming solar radiation and which were accompanied by a small reduction in solar output during the period. These factors, along with natural decades-long oscillations in the climate system, may have masked a portion of the greenhouse warming. (However, climatologists point out that these natural climate cycles are expected to add to greenhouse warming in the future when the oscillations eventually reverse direction.) For these reasons many scientists believe that it is an error to call this slowdown in detectable surface warming a "hiatus" or a "pause."

Theoretical Climate Models

Theoretical models of Earth's climate system can be used to investigate the response of climate to external radiative forcing as well as its own internal variability. Two or more models that focus on different physical processes may be coupled or linked together through a common feature, such as geographic location. Climate models vary considerably in their degree of complexity. The simplest models of energy balance describe Earth's surface as a globally uniform layer whose temperature is determined by a balance of incoming and outgoing shortwave and longwave radiation. These simple models may also consider the effects of greenhouse gases. At the other end of the spectrum are fully coupled, three-dimensional, global climate models. These are complex models that solve for radiative balance; for laws of motion governing the atmosphere, ocean, and ice; and for exchanges of energy and momentum within and between the different components of the climate. In some cases, theoretical climate models also include an interactive representation of Earth's biosphere and carbon cycle.

Even the most-detailed climate models cannot resolve all the processes that are important in the atmosphere and ocean. Most climate models are designed to gauge the behaviour of a number of physical variables over space and time, and they often artificially divide Earth's surface into a grid of many equal-sized "cells." Each cell may neatly correspond to some physical process (such as summer near-surface air temperature) or other variable (such as land-use type), and it may be assigned a relatively straightforward value. So-called "sub-grid-scale" processes, such as those of clouds, are too small to be captured by the relatively coarse spacing of the individual grid cells. Instead, such processes must be represented through a statistical process that relates the properties of the atmosphere and ocean. For example, the average fraction of cloud cover over a hypothetical "grid box" (that is, a representative volume of air or water in the model) can be estimated from the average relative humidity and the vertical temperature profile of the grid cell. Variations in the behaviour of different coupled climate models arise in large part from differences in the ways sub-grid-scale processes are mathematically expressed.

Despite these required simplifications, many theoretical climate models perform remarkably well when reproducing basic features of the atmosphere, such as the behaviour of midlatitude jet streams or Hadley cell circulation. The models also adequately reproduce important features of the oceans, such as the Gulf Stream. In addition, models are becoming better able to reproduce the main patterns of internal climate variability, such as those of El Niño/Southern Oscillation

(ENSO). Consequently, periodically recurring events—such as ENSO and other interactions between the atmosphere and ocean currents—are being modeled with growing confidence.

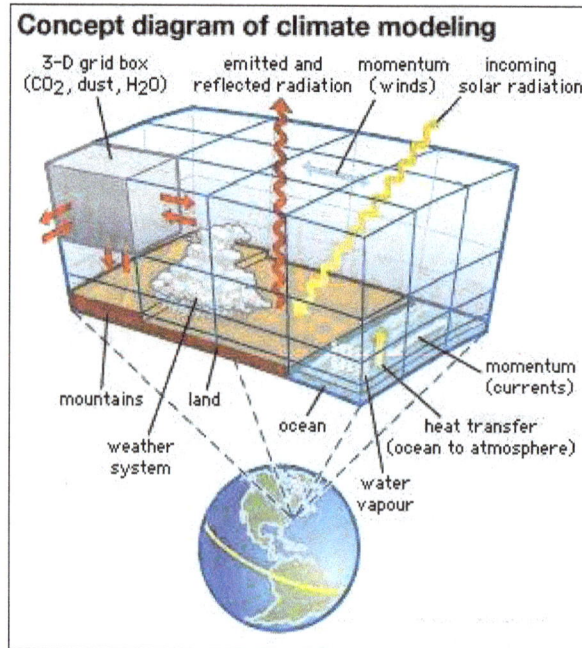

To understand and explain the complex behaviour of Earth's climate, modern climate models incorporate several variables that stand in for materials passing through Earth's atmosphere and oceans and the forces that affect them.

Climate models have been tested in their ability to reproduce observed changes in response to radiative forcing. In 1988 a team at NASA's Goddard Institute for Space Studies in New York City used a fairly primitive climate model to predict warming patterns that might occur in response to three different scenarios of anthropogenic radiative forcing. Warming patterns were forecast for subsequent decades. Of the three scenarios, the middle one, which corresponds most closely to actual historical carbon emissions, comes closest to matching the observed warming of roughly 0.5 °C (0.9 °F) that has taken place since then. The NASA team also used a climate model to successfully predict that global mean surface temperatures would cool by about 0.5 °C for one to two years after the 1991 eruption of Mount Pinatubo in the Philippines.

More recently, so-called "detection and attribution" studies have been performed. These studies compare predicted changes in near-surface air temperature and other climate variables with patterns of change that have been observed for the past one to two centuries. The simulations have shown that the observed patterns of warming of Earth's surface and upper oceans, as well as changes in other climate phenomena such as prevailing winds and precipitation patterns, are consistent with the effects of an anthropogenic influence predicted by the climate models. In addition, climate model simulations have shown success in reproducing the magnitude and the spatial pattern of cooling in the Northern Hemisphere between roughly 1400 and 1850—during the Little Ice Age, which appears to have resulted from a combination of lowered solar output and heightened explosive volcanic activity.

Potential effects of Global Warming

The path of future climate change will depend on what courses of action are taken by society—in particular the emission of greenhouse gases from the burning of fossil fuels. A range of alternative emissions scenarios known as representative concentration pathways (RCPs) were proposed by the IPCC in the Fifth Assessment Report (AR5), which was published in 2014, to examine potential future climate changes. The scenarios depend on various assumptions concerning future rates of human population growth, economic development, energydemand, technological advancement, and other factors. Unlike the scenarios used in previous IPCC assessments, the AR5 RCPs explicitly account for climate change mitigation efforts.

Projected range of sea-level rise by climate change scenario.		
Scenario	Temperature change (°C) in 2090–99 relative to 1980–99	Sea-level rise (m) in 2090–99 relative to 1980–99
B1	1.1–2.9	0.18–0.38
A1T	1.4–3.8	0.20–0.45
B2	1.4–3.8	0.20–0.43
A1B	1.7–4.4	0.21–0.48
A2	2.0–5.4	0.23–0.51
A1Fl	2.4–6.4	0.26–0.59

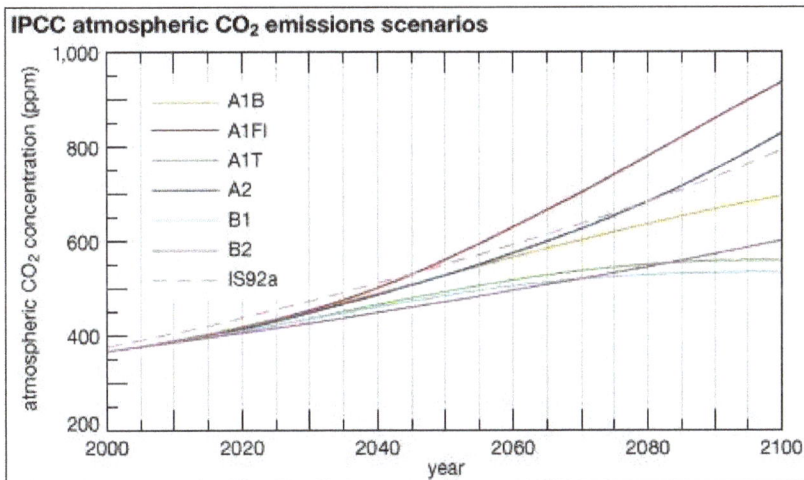

Carbon dioxide: Global warming scenarios Graph of the predicted increase in the concentration of carbon dioxide (CO_2) in Earth's atmosphere according to a series of climate change scenarios that assume different levels of economic development, population growth, and fossil fuel use.

The results of each scenario in the IPCC's Fourth Assessment Report are depicted in the graph.

The AR5 scenario with the smallest increases in greenhouse gases is RCP 2.6, which denotes the net radiative forcing by 2100 in watts per square metre (a doubling of CO_2 concentrations from preindustrial values of 280 ppm to 560 ppm represents roughly 3.7 watts per square metre). RCP 2.6 assumes substantial improvements in energy efficiency, a rapid transition away from fossil fuel energy, and a global population that peaks at roughly nine billion people in the 21st century. In that scenario CO_2 concentrations remain below 450 ppm and actually falltoward the end of the century (to about 420 ppm) as a result of widespread deployment of carbon-capture technology.

Scenario RCP 8.5, by contrast, might be described as "business as usual." It reflects the assumption of an energy-intensive global economy, high population growth, and a reduced rate of technological development. CO_2 concentrations are more than three times greater than preindustrial levels (roughly 936 ppm) by 2100 and continue to grow thereafter. RCP 4.5 and RCP 6.0 envision intermediate policy choices, resulting in stabilization by 2100 of CO_2 concentrations at 538 and 670 ppm, respectively. In all those scenarios, the cooling effect of industrial pollutants such as sulfate particulates, which have masked some of the past century's warming, is assumed to decline to near zero by 2100 because of policies restricting their industrial production.

Simulations of Future Climate Change

The differences between the various simulations arise from disparities between the various climate models used and from assumptions made by each emission scenario. For example, best estimates of the predicted increases in global surface temperature between the years 2000 and 2100 range from about 0.3 to 4.8 °C (0.5 to 8.6 °F), depending on which emission scenario is assumed and which climate model is used. Relative to preindustrial temperatures, these estimates reflect an overall warming of the globe of 1.4 to 5.0 °C (2.5 to 9.0 °F). These projections are conservative in that they do not take into account potential positive carbon cycle feedbacks. Only the lower-end emissions scenario RCP 2.6 has a reasonable chance (roughly 50 percent) of holding additional global surface warming by 2100 to less than 2.0 °C (3.6 °F)—a level considered by many scientists to be the threshold above which pervasive and extreme climatic effects will occur.

Patterns of Warming

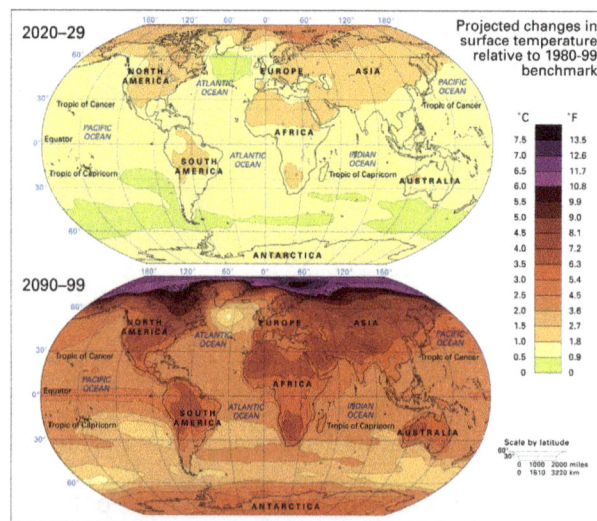

Projected changes in mean surface temperatures by the late 21st century according to the A1B climate change scenario. All values for the period 2090–99 are shown relative to the mean temperature values for the period 1980–99.

The greatest increase in near-surface air temperature is projected to occur over the polar region of the Northern Hemisphere because of the melting of sea ice and the associated reduction in surface albedo. Greater warming is predicted over land areas than over the ocean. Largely due to the delayed warming of the oceans and their greater specific heat, the Northern Hemisphere—with less than 40 percent of its surface area covered by water—is expected to warm faster than the Southern

Hemisphere. Some of the regional variation in predicted warming is expected to arise from changes to wind patterns and ocean currents in response to surface warming. For example, the warming of the region of the North Atlantic Ocean just south of Greenland is expected to be slight. This anomaly is projected to arise from a weakening of warm northward ocean currents combined with a shift in the jet stream that will bring colder polar air masses to the region.

Precipitation Patterns

The climate changes associated with global warming are also projected to lead to changes in precipitation patterns across the globe. Increased precipitation is predicted in the polar and subpolar regions, whereas decreased precipitation is projected for the middle latitudes of both hemispheres as a result of the expected poleward shift in the jet streams. Whereas precipitation near the Equator is predicted to increase, it is thought that rainfall in the subtropics will decrease. Both phenomena are associated with a forecasted strengthening of the tropical Hadley cell pattern of atmospheric circulation.

Changes in precipitation patterns are expected to increase the chances of both drought and flood conditions in many areas. Decreased summer precipitation in North America, Europe, and Africa, combined with greater rates of evaporation due to warming surface temperatures, is projected to lead to decreased soil moisture and drought in many regions. Furthermore, since anthropogenic climate change will likely lead to a more vigorous hydrologic cycle with greater rates of both evaporation and precipitation, there will be a greater probability for intense precipitation and flooding in many regions.

Regional Predictions

Regional predictions of future climate change remain limited by uncertainties in how the precise patterns of atmospheric winds and ocean currents will vary with increased surface warming. For example, some uncertainty remains in how the frequency and magnitude of El Niño/Southern Oscillation (ENSO) events will adjust to climate change. Since ENSO is one of the most prominent sources of interannual variations in regional patterns of precipitation and temperature, any uncertainty in how it will change implies a corresponding uncertainty in certain regional patterns of climate change. For example, increased El Niño activity would likely lead to more winter precipitation in some regions, such as the desert southwest of the United States. This might offset the drought predicted for those regions, but at the same time it might lead to less precipitation in other regions. Rising winter precipitation in the desert southwest of the United States might exacerbate drought conditions in locations as far away as South Africa.

Ice Melt and Sea Level Rise

A warming climate holds important implications for other aspects of the global environment. Because of the slow process of heat diffusion in water, the world's oceans are likely to continue to warm for several centuries in response to increases in greenhouse concentrations that have taken place so far. The combination of seawater's thermal expansion associated with this warming and the melting of mountain glaciers is predicted to lead to an increase in global sea level of 0.45–0.82 metre (1.4–2.7 feet) by 2100 under the RCP 8.5 emissions scenario. However, the actual rise in sea level could be considerably greater than this. It is probable that the continued warming of Greenland will cause its ice sheet to melt at accelerated rates. In addition, this level of surface warming may also melt the ice sheet of West Antarctica. Paleoclimatic evidence suggests that an additional

2 °C (3.6 °F) of warming could lead to the ultimate destruction of the Greenland Ice Sheet, an event that would add another 5 to 6 metres (16 to 20 feet) to predicted sea level rise. Such an increase would submerge a substantial number of islands and lowland regions. Coastal lowland regions vulnerable to sea level rise include substantial parts of the U.S. Gulf Coast and Eastern Seaboard (including roughly the lower third of Florida), much of the Netherlands and Belgium (two of the European Low Countries), and heavily populated tropical areas such as Bangladesh. In addition, many of the world's major cities—such as Tokyo, New York, Mumbai, Shanghai, and Dhaka—are located in lowland regions vulnerable to rising sea levels. With the loss of the West Antarctic ice sheet, additional sea level rise would approach 10.5 metres (34 feet).

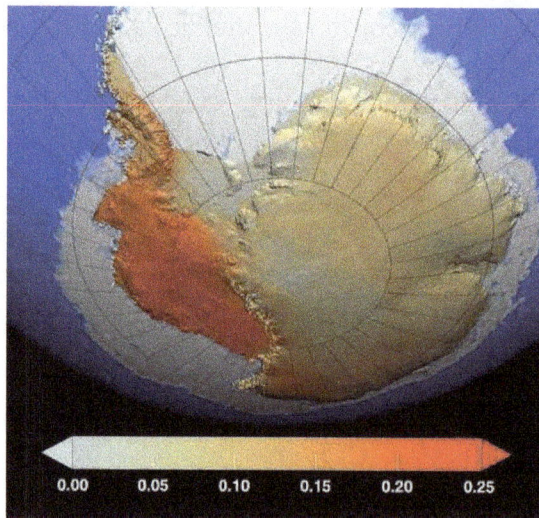

Figure show NASA image showing locations on Antarctica where temperatures had increased between 1959 and 2009. Red represents areas where temperatures had increased the most over the period, particularly in West Antarctica, while dark blue represents areas with a lesser degree of warming. Temperature changes are measured in degrees Celsius.

While the current generation of models predicts that such global sea level changes might take several centuries to occur, it is possible that the rate could accelerate as a result of processes that tend to hasten the collapse of ice sheets. One such process is the development of moulins—large vertical shafts in the ice that allow surface meltwater to penetrate to the base of the ice sheet. A second process involves the vast ice shelves off Antarctica that buttress the grounded continental ice sheet of Antarctica's interior. If those ice shelves collapse, the continental ice sheet could become unstable, slide rapidly toward the ocean, and melt, thereby further increasing mean sea level. Thus far, neither process has been incorporated into the theoretical models used to predict sea level rise.

Ocean Circulation Changes

Another possible consequence of global warming is a decrease in the global ocean circulation system known as the "thermohaline circulation" or "great ocean conveyor belt." This system involves the sinking of cold saline waters in the subpolar regions of the oceans, an action that helps to drive warmer surface waters poleward from the subtropics. As a result of this process, a warming influence is carried to Iceland and the coastal regions of Europe that moderates the climate in those regions. Some scientists believe that global warming could shut down this ocean current system by creating an influx of fresh water from melting ice sheets and glaciers into the subpolar North Atlantic Ocean. Since fresh water is less dense than saline water, a significant intrusion of fresh water would lower the density of the surface waters and thus inhibit the sinking motion that drives the large-scale thermohaline circulation. It has also been speculated that, as a consequence of large-scale surface warming, such changes could even trigger colder conditions in regions surrounding the North Atlantic. Experiments with modern climate models suggest that such an event would be unlikely. Instead, a moderate weakening of the thermohaline circulation might occur that would lead to a dampening of surface warming—rather than actual cooling—in the higher latitudes of the North Atlantic Ocean.

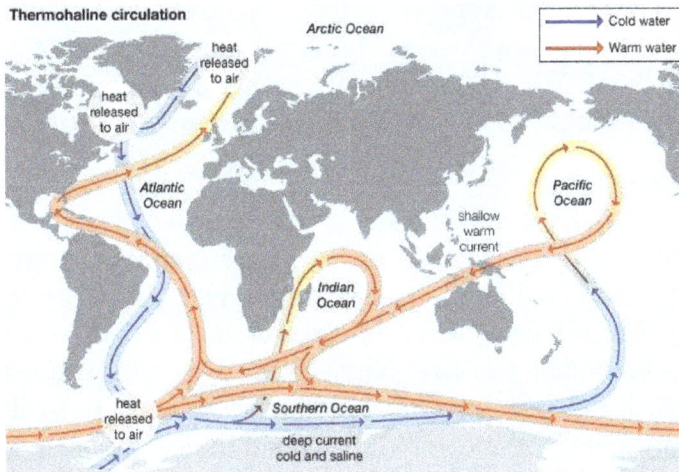

Figure show thermohaline circulation transports and mixes the water of the oceans. In the process it transports heat, which influences regional climate patterns. The density of seawater is determined by the temperature and salinity of a volume of seawater at a particular location. The difference in density between one location and another drives the thermohaline circulation.

Tropical Cyclones

One of the more controversial topics in the science of climate change involves the impact of global warming on tropical cyclone activity. It appears likely that rising tropical ocean temperatures associated with global warming will lead to an increase in the intensity (and the associated destructive potential) of tropical cyclones. In the Atlantic a close relationship has been observed between rising ocean temperatures and a rise in the strength of hurricanes. Trends in the intensities of tropical cyclones in other regions, such as in the tropical Pacific and Indian oceans, are more uncertain due to a paucity of reliable long-term measurements.

While the warming of oceans favours increased tropical cyclone intensities, it is unclear to what extent rising temperatures affect the number of tropical cyclones that occur each year. Other factors, such as wind shear, could play a role. If climate change increases the amount of wind shear—a factor that discourages the formation of tropical cyclones—in regions where such storms tend to form, it might partially mitigate the impact of warmer temperatures. On the other hand, changes in atmospheric winds are themselves uncertain—because of, for example, uncertainties in how climate change will affect ENSO.

Environmental Consequences of Global Warming

Global warming and climate change have the potential to alter biological systems. More specifically, changes to near-surface air temperatures will likely influence ecosystemfunctioning and thus the biodiversity of plants, animals, and other forms of life. The current geographic ranges of plant and animal species have been established by adaptation to long-term seasonal climate patterns. As global warming alters these patterns on timescales considerably shorter than those that arose in the past from natural climate variability, relatively sudden climatic changes may challenge the natural adaptive capacity of many species.

A large fraction of plant and animal species are likely to be at an increased risk of extinction if global average surface temperatures rise another 1.5 to 2.5 °C (2.7 to 4.5 °F) by the year 2100. Species loss estimates climb to as much as 40 percent for a warming in excess of 4.5 °C (8.1 °F)—a level that could be reached in the IPCC's higher emissions scenarios. A 40 percent extinction rate would likely lead to major changes in the food webs within ecosystems and have a destructive impact on ecosystem function.

Surface warming in temperate regions is likely to lead changes in various seasonal processes—for instance, earlier leaf production by trees, earlier greening of vegetation, altered timing of egg laying and hatching, and shifts in the seasonal migration patterns of birds, fishes, and other migratory animals. In high-latitude ecosystems, changes in the seasonal patterns of sea ice threaten predators such as polar bears and walruses; both species rely on broken sea ice for their hunting activities. Also in the high latitudes, a combination of warming waters, decreased sea ice, and changes in ocean salinity and circulation is likely to lead to reductions or redistributions in populations of algae and plankton. As a result, fish and other organisms that forage upon algae and plankton may be threatened. On land, rising temperatures and changes in precipitation patterns and drought frequencies are likely to alter patterns of disturbance by fires and pests.

Numerous ecologists, conservation biologists, and other scientists studying climate warn that rising surface temperatures will bring about an increased extinction risk. In 2015 one study that examined 130 extinction models developed in previous studies predicted that 5.2 percent of species would be lost with a rise in average temperatures of 2 °C (3.6 °F) above temperature benchmarks from before the onset of the Industrial Revolution. The study also predicted that 16 percent of Earth's species would be lost if surface warming increased to about 4.3 °C (7.7 °F) above preindustrial temperature benchmarks.

Other likely impacts on the environment include the destruction of many coastal wetlands, salt marshes, and mangrove swamps as a result of rising sea levels and the loss of certain rare and fragile habitats that are often home to specialist species that are unable to thrive in other environments. For example, certain amphibians limited to isolated tropical cloud forests either have become extinct already or are under serious threat of extinction. Cloud forests—tropical forests that depend on persistent condensation of moisture in the air—are disappearing as optimal condensation levels move to higher elevations in response to warming temperatures in the lower atmosphere.

In many cases a combination of stresses caused by climate change as well as human activity represents a considerably greater threat than either climatic stresses or nonclimatic stresses alone. A particularly important example is coral reefs, which contain much of the ocean's biodiversity. Rising ocean temperatures increase the tendency for coral bleaching (a condition where zooxanthellae, or yellow-green algae, living in symbiosis with coral either lose their pigments or abandon the coral polyps altogether), and they also raise the likelihood of greater physical damage by progressively more destructive tropical cyclones. In many areas coral is also under stress from increased ocean acidification, marine pollution, runoff from agricultural fertilizer, and physical damage by boat anchors and dredging.

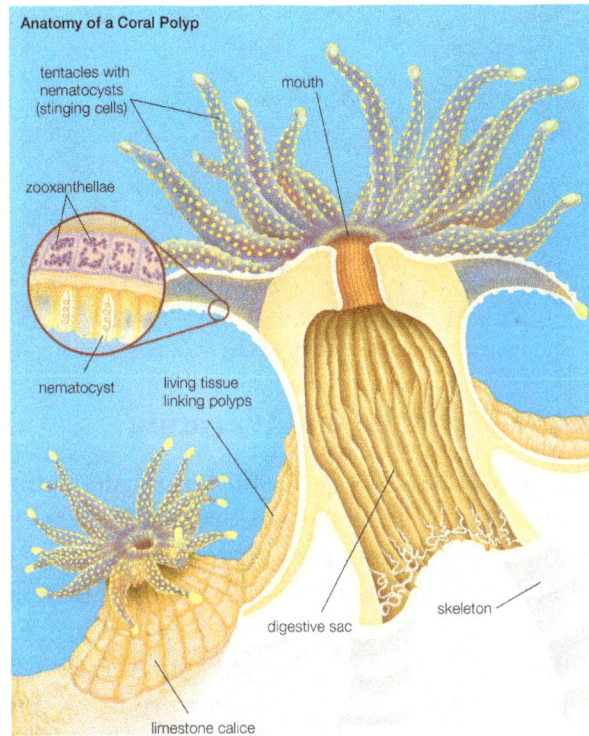

Cross section of a generalized coral polyp.

Another example of how climate and nonclimatic stresses combine is illustrated by the threat to migratory animals. As these animals attempt to relocate to regions with more favourable climate conditions, they are likely to encounter impediments such as highways, walls, artificial waterways, and other man-made structures.

Warmer temperatures are also likely to affect the spread of infectious diseases, since the geographic ranges of carriers, such as insects and rodents, are often limited by climatic conditions. Warmer winter conditions in New York in 1999, for example, appear to have facilitated an outbreak of West Nile virus, whereas the lack of killing frosts in New Orleansduring the early 1990s led to an explosion of disease-carrying mosquitoes and cockroaches. Warmer winters in the Korean peninsula and southern Europe have allowed the spread of the *Anopheles* mosquito, which carries the malaria parasite, whereas warmer conditions in Scandinavia in recent years have allowed for the northward advance of encephalitis.

In the southwestern United States, alternations between drought and flooding related in part to the ENSO phenomenon have created conditions favourable for the spread of hantavirusesby rodents. The spread of mosquito-borne Rift Valley fever in equatorial East Africa has also been related to wet conditions in the region associated with ENSO. Severe weather conditions conducive to rodents or insects have been implicated in infectious disease outbreaks—for instance, the outbreaks of cholera and leptospirosis that occurred after Hurricane Mitch struck Central America in 1998. Global warming could therefore affect the spread of infectious disease through its influence on ENSO or on severe weather conditions.

Anopheles mosquito, carrier of the malarial parasite.

Socioeconomic Consequences of Global Warming

Socioeconomic impacts of global warming could be substantial, depending on the actual temperature increases over the next century. Models predict that a net global warming of 1 to 3 °C (1.8 to 5.4 °F) beyond the late 20th-century global average would produce economic losses in some regions (particularly the tropics and high latitudes) and economic benefits in others. For warming beyond those levels, benefits would tend to decline and costs increase. For warming in excess of 4 °C (7.2 °F), models predict that costs will exceed benefits on average, with global mean economic losses estimated between 1 and 5 percent of gross domestic product. Substantial disruptions could be expected under those conditions, specifically in the areas of agriculture, food and forest products, water and energy supply, and human health.

Agricultural productivity might increase modestly in temperate regions for some crops in response to a local warming of 1–3 °C (1.8–5.4 °F), but productivity will generally decrease with further warming. For tropical and subtropical regions, models predict decreases in crop productivity for even small increases in local warming. In some cases, adaptations such as altered planting practices are projected to ameliorate losses in productivity for modest amounts of warming. An increased incidence of drought and flood events would likely lead to further decreases in agricultural productivity and to decreases in livestock production, particularly among subsistence farmers in tropical regions. In regions such as the African Sahel, decreases in agricultural productivity have already been observed as a result of shortened growing seasons, which in turn have occurred as a result of warmer and drier climatic conditions. In other regions, changes in agricultural practice, such as planting crops earlier in the growing season, have been undertaken. The warming of oceans is predicted to have an adverse impact on commercial fisheries by changing the distribution and productivity of various fish species, whereas commercial timber productivity may increase globally with modest warming.

Water resources are likely to be affected substantially by global warming. At current rates of warming, a 10–40 percent increase in average surface runoff and water availability has been projected in higher latitudes and in certain wet regions in the tropics by the middle of the 21st century, while decreases of similar magnitude are expected in other parts of the tropics and in the dry regions in the subtropics. This would be particularly severe during the summer season. In many cases water availability is already decreasing or expected to decrease in regions that have been stressed for water resources since the turn of the 21st century. Such regions as the African Sahel, western North America, southern Africa, the Middle East, and western Australiacontinue to be particularly vulnerable. In these regions drought is projected to increase in both magnitude and extent, which would bring about adverse effects on agriculture and livestock raising. Earlier and increased spring runoff is already being observed in western North America and other temperate regions served by glacial or snow-fed streams and rivers. Fresh watercurrently stored by mountain glaciers and snow in both the tropics and extratropics is also projected to decline and thus reduce the availability of fresh water for more than 15 percent of the world's population. It is also likely that warming temperatures, through their impact on biological activity in lakes and rivers, may have an adverse impact on water quality, further diminishing access to safe water sources for drinking or farming. For example, warmer waters favour an increased frequency of nuisance algal blooms, which can pose health risks to humans. Risk-management procedures have already been taken by some countries in response to expected changes in water availability.

Energy availability and use could be affected in at least two distinct ways by rising surface temperatures. In general, warmer conditions would favour an increased demand for air-conditioning; however, this would be at least partially offset by decreased demand for winterheating in temperate regions. Energy generation that requires water either directly, as in hydroelectric power, or indirectly, as in steam turbines used in coal-fired power plants or in cooling towers used in nuclear power plants, may become more difficult in regions with reduced water supplies.

As, it is expected that human health will be further stressed under global warming conditions by potential increases in the spread of infectious diseases. Declines in overall human health might occur with increases in the levels of malnutrition due to disruptions in food production and by increases in the incidence of afflictions. Such afflictions could include diarrhea, cardiorespiratory illness, and allergic reactions in the midlatitudes of the Northern Hemisphere as a result of rising levels of pollen. Rising heat-related mortality, such as that observed in response to the 2003 European heat wave, might occur in many regions, especially in impoverished areas where air-conditioning is not generally available.

The economic infrastructure of most countries is predicted to be severely strained by global warming and climate change. Poor countries and communities with limited adaptive capacities are likely to be disproportionately affected. Projected increases in the incidence of severe weather, heavy flooding, and wildfires associated with reduced summer ground moisture in many regions will threaten homes, dams, transportation networks and other facets of human infrastructure. In high-latitude and mountain regions, melting permafrost is likely to lead to ground instability or rock avalanches, further threatening structures in those regions. Rising sea levels and the increased potential for severe tropical cyclones represent a heightened threat to coastal communities throughout the world. It has been estimated that an additional warming of 1–3 °C (1.8–5.4 °F) beyond the late 20th-century global average would threaten millions more people with the risk of annual flooding. People in the densely populated, poor, low-lying regions of Africa, Asia, and tropical islands would be the most vulnerable, given their limited adaptive capacity. In addition, certain regions in developed countries, such as the Low Countries of Europe and the Eastern Seaboard and Gulf Coast of the United States, would also be vulnerable to the effects of rising sea levels. Adaptive steps are already being taken by some governments to reduce the threat of increased coastal vulnerability through the construction of dams and drainage works.

Greenhouse Effect

The greenhouse effect is the process by which radiation from a planet's atmosphere warms the planet's surface to a temperature above what it would be without this atmosphere.

Radiatively active gases (i.e., greenhouse gases) in a planet's atmosphere radiate energy in all directions. Part of this radiation is directed towards the surface, warming it. The intensity of the downward radiation – that is, the strength of the greenhouse effect – will depend on the atmosphere's temperature and on the amount of greenhouse gases that the atmosphere contains.

Earth's natural greenhouse effect is critical to supporting life, and initially was a precursor to life moving out of the ocean onto land. Human activities, however, mainly the burning of fossil fuels and clearcutting of forests, have accelerated the greenhouse effect and caused global warming.

The planet Venus experienced runaway greenhouse effect, resulting in an atmosphere which is 96% carbon dioxide, with surface atmospheric pressure roughly the same as found 900 m (3,000 ft) underwater on Earth. Venus may have had water oceans, but they would have boiled off as the mean surface temperature rose to 735 K (462 °C; 863 °F).

The term "greenhouse effect" continues to see use in scientific circles and the media despite being a slight misnomer, as an atmosphere reduces radiative heat loss while a greenhouse blocks convective heat loss. The result, however, is an increase in temperature in both cases.

The existence of the greenhouse effect was argued for by Joseph Fourier in 1824. The argument and the evidence were further strengthened by Claude Pouillet in 1827 and 1838 and reasoned from experimental observations by Eunice Newton Foote in 1856. John Tyndall expanded her work in 1859 by measuring radiative properties of a wider spectrum of greenhouse gases. The effect was more fully quantified by Svante Arrhenius in 1896, who made the first quantitative prediction of global warming due to a hypothetical doubling of atmospheric carbon dioxide. However, the term "greenhouse" was not used to refer to this effect by any of these scientists; the term was first used in this way by Nils Gustaf Ekholm in 1901.

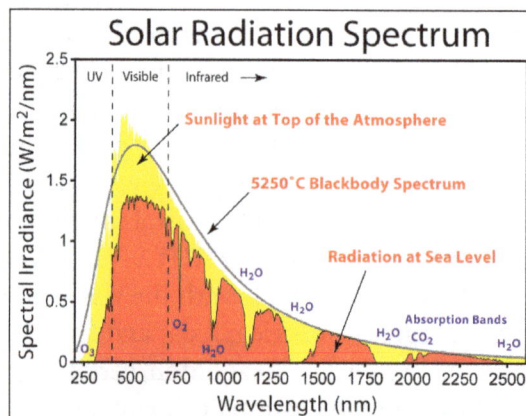

The solar radiation spectrum for direct light at both the top of Earth's atmosphere and at sea level.

Earth receives energy from the Sun in the form of ultraviolet, visible, and near-infrared radiation. About 26% of the incoming solar energy is reflected to space by the atmosphere and clouds, and

19% is absorbed by the atmosphere and clouds. Most of the remaining energy is absorbed at the surface of Earth. Because the Earth's surface is colder than the Sun, it radiates at wavelengths that are much longer than the wavelengths that were absorbed. Most of this thermal radiation is absorbed by the atmosphere and warms it. The atmosphere also gains heat by sensible and latent heat fluxes from the surface. The atmosphere radiates energy both upwards and downwards; the part radiated downwards is absorbed by the surface of Earth. This leads to a higher equilibrium temperature than if the atmosphere did not radiate.

An ideal thermally conductive blackbody at the same distance from the Sun as Earth would have a temperature of about 5.3 °C (41.5 °F). However, because Earth reflects about 30% of the incoming sunlight, this idealized planet's effective temperature (the temperature of a blackbody that would emit the same amount of radiation) would be about –18 °C (0 °F). The surface temperature of this hypothetical planet is 33 °C (59 °F) below Earth's actual surface temperature of approximately 14 °C (57 °F). The greenhouse effect is the contribution of greenhouse gases to this difference.

Mechanism

The basic mechanism can be qualified in a number of ways, none of which affect the fundamental process. The atmosphere near the surface is largely opaque to thermal radiation (with important exceptions for "window" bands), and most heat loss from the surface is by sensible heat and latent heat transport. Radiative energy losses become increasingly important higher in the atmosphere, largely because of the decreasing concentration of water vapor, an important greenhouse gas. It is more realistic to think of the greenhouse effect as applying to a layer in the mid-troposphere, which is effectively coupled to the surface by a lapse rate. The simple picture also assumes a steady state, but in the real world, the diurnal cycle as well as the seasonal cycle and weather disturbances complicate matters. Solar heating applies only during daytime. During the night, the atmosphere cools somewhat, but not greatly, because its emissivity is low. Diurnal temperature changes decrease with height in the atmosphere.

Within the region where radiative effects are important, the description given by the idealized greenhouse model becomes realistic. Earth's surface, warmed to a temperature around 255 K, radiates long-wavelength, infrared heat in the range of 4–100 μm. At these wavelengths, greenhouse gases that were largely transparent to incoming solar radiation are more absorbent. Each layer of atmosphere with greenhouses gases absorbs some of the heat being radiated upwards from lower layers. It reradiates in all directions, both upwards and downwards; in equilibrium (by definition) the same amount as it has absorbed. This results in more warmth below. Increasing the concentration of the gases increases the amount of absorption and reradiation, and thereby further warms the layers and ultimately the surface below.

Greenhouse gases—including most diatomic gases with two different atoms (such as carbon monoxide, CO) and all gases with three or more atoms—are able to absorb and emit infrared radiation. Though more than 99% of the dry atmosphere is IR transparent (because the main constituents—N_2, O_2, and Ar—are not able to directly absorb or emit infrared radiation), intermolecular collisions cause the energy absorbed and emitted by the greenhouse gases to be shared with the other, non-IR-active, gases.

Greenhouse Gases

By their percentage contribution to the greenhouse effect on Earth the four major gases are:

- Water vapor, 36–70%,

- Carbon dioxide, 9–26%,

- Methane, 4–9%,

- Ozone, 3–7%.

Atmospheric gases only absorb some wavelengths of energy but are transparent to others. The absorption patterns of water vapor (blue peaks) and carbon dioxide (pink peaks) overlap in some wavelengths. Carbon dioxide is not as strong a greenhouse gas as water vapor, but it absorbs energy in longer wavelengths (12–15 micrometers) that water vapor does not, partially closing the "window" through which heat radiated by the surface would normally escape to space.

It is not possible to assign a specific percentage to each gas because the absorption and emission bands of the gases overlap (hence the ranges given above). Clouds also absorb and emit infrared radiation and thus affect the radiative properties of the atmosphere.

Role in Climate Change

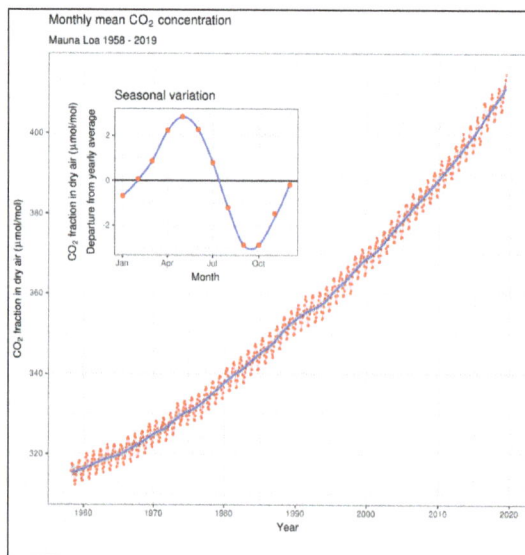

The Keeling Curve of atmospheric CO_2 concentrations measured at Mauna Loa Observatory.

Strengthening of the greenhouse effect through human activities is known as the enhanced (or anthropogenic) greenhouse effect. This increase in radiative forcing from human activity is attributable mainly to increased atmospheric carbon dioxide levels. According to the latest Assessment Report from the Intergovernmental Panel on Climate Change, "atmospheric concentrations of carbon dioxide, methane and nitrous oxide are unprecedented in at least the last 800,000 years. Their effects, together with those of other anthropogenic drivers, have been detected throughout the climate system and are extremely likely to have been the dominant cause of the observed warming since the mid-20th century".

CO_2 is produced by fossil fuel burning and other activities such as cement production and tropical deforestation. Measurements of CO_2 from the Mauna Loa observatory show that concentrations have increased from about 313 parts per million (ppm) in 1960 to about 389 ppm in 2010. It reached the 400 ppm milestone on May 9, 2013. The current observed amount of CO_2 exceeds the geological record maxima (~300 ppm) from ice core data. The effect of combustion-produced carbon dioxide on the global climate, a special case of the greenhouse effect first described in 1896 by Svante Arrhenius, has also been called the Callendar effect.

Over the past 800,000 years, ice core data shows that carbon dioxide has varied from values as low as 180 ppm to the pre-industrial level of 270 ppm. Paleoclimatologists consider variations in carbon dioxide concentration to be a fundamental factor influencing climate variations over this time scale.

Real Greenhouses

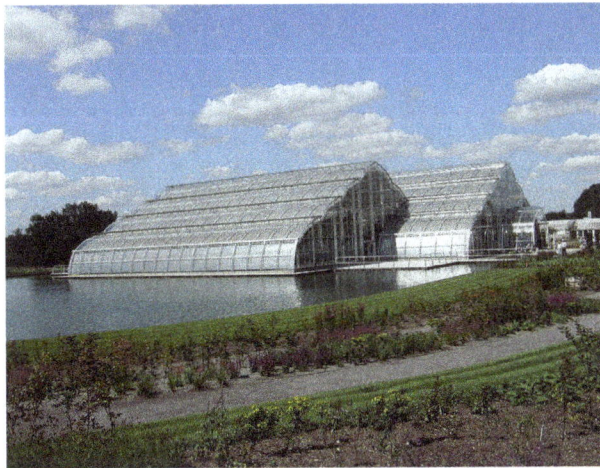

A modern Greenhouse in RHS Wisley.

The "greenhouse effect" of the atmosphere is named by analogy to greenhouses which become warmer in sunlight. However, a greenhouse is not primarily warmed by the "greenhouse effect". "Greenhouse effect" is actually a misnomer since heating in the usual greenhouse is due to the reduction of convection, while the "greenhouse effect" works by preventing absorbed heat from leaving the structure through radiative transfer.

A greenhouse is built of any material that passes sunlight: usually glass or plastic. The sun warms the ground and contents inside just like the outside, and these then warm the air. Outside, the warm air near the surface rises and mixes with cooler air aloft, keeping the temperature lower than inside, where the air continues to heat up because it is confined within the greenhouse. This can be demonstrated by opening a small window near the roof of a greenhouse: the temperature will drop considerably. It was demonstrated experimentally that a (not heated) "greenhouse" with a cover of rock salt (which is transparent to infrared) heats up an enclosure similarly to one with a glass cover. Thus greenhouses work primarily by preventing convective cooling.

Heated greenhouses are yet another matter: as they have an internal source of heating, it is desirable to minimise the amount of heat leaking out by radiative cooling. This can be done through the use of adequate glazing.

Related Effects

Anti-greenhouse Effect

The anti-greenhouse effect is a mechanism similar and symmetrical to the greenhouse effect: in the greenhouse effect, the atmosphere lets radiation in while not letting thermal radiation out, thus warming the body surface; in the anti-greenhouse effect, the atmosphere keeps radiation out while letting thermal radiation out, which lowers the equilibrium surface temperature. Such an effect has been proposed for Saturn's moon Titan.

Runaway Greenhouse Effect

A runaway greenhouse effect occurs if positive feedbacks lead to the evaporation of all greenhouse gases into the atmosphere. A runaway greenhouse effect involving carbon dioxide and water vapor has long ago been hypothesized to have occurred on Venus, this idea is still largely accepted.

Bodies other than Earth

The 'greenhouse effect' on Venus is particularly large for several reasons:

- It is nearer to the Sun than Earth by about 30%.

- It has a much weaker magnetic field meaning less protection from solar radiation and heating effects.

- Its very dense atmosphere consists mainly of carbon dioxide.

Venus experienced a runaway greenhouse in the past, and we expect that Earth will in about 2 billion years as solar luminosity increases.

In complete contast the mean temparature on Mars is very cold at -63 °C (-82 °F). This is despite having over 95% atmospheric CO_2, almost the same as Venus's atmosphere, but at a much lower pressure than the atmospheres on Earth and Venus. Mars is further from the Sun than Earth but still receives about 44% of the Sun's heat [approx 500 W/m²] when compared to Earth. Also any atmospheric or surface heating from solar flares and cosmic radiation affects Mars as it has no magnetic field.

Titan has an anti-greenhouse effect, in that its atmosphere absorbs solar radiation but is relatively transparent to outgoing infrared radiation.

Pluto is also colder than would be expected because it is cooled by evaporation of nitrogen.

Climate Change Adaptation

Effects of Global Warming

The projected effects for the environment and for civilization are numerous and varied. The main effect is an increasing global average temperature. The average surface temperature could increase by 3 to 10 degrees Fahrenheit (approximately 1.67 to 5.56 °C) by the end of the century if carbon emissions aren't reduced. This causes a variety of secondary effects, namely, changes in patterns of precipitation, rising sea levels, altered patterns of agriculture, increased extreme weather events, the expansion of the range of tropical diseases, and the opening of new marine trade routes.

Potential effects include sea level rise of 110 to 770 mm (0.36 to 2.5 feet) between 1990 and 2100, repercussions to agriculture, possible slowing of the thermohaline circulation, reductions in the ozone layer, increased intensity and frequency of extreme weather events, lowering of ocean pH, and the spread of tropical diseases such as malaria and dengue fever.

A summary of probable effects and recent understanding can be found in the report made for the IPCC Third Assessment Report by Working Group II. The 2007 contribution of Working Group II detailing the impacts of global warming for the IPCC Fourth Assessment Report has been summarized for policymakers.

Adaptation is handicapped by uncertainty over the effects of global warming on specific locations such as the Southwestern United States or phenomena such as the Indian monsoon predicted to increase in frequency and intensity.

International Adaptation Finance

The United Nations Framework Convention on Climate Change, under Article 11, incorporates a financial mechanism to developing country parties to support them with adaptation. Until 2009, three funds existed under the UNFCCC financial mechanism. The Special Climate Change Fund (SCCF) and the Least Developed Countries Fund (LDCF) are administered by the Global Environmental Facility. The Adaptation Fund was established a result of negotiations during COP15 and COP16 and is administered by its own Secretariat. Initially, when the Kyoto Protocol was in operation, the Adaptation Fund was financed by a 2% levy on the Clean Development Mechanism (CDM).

At the 15th Conference of the Parties to the UNFCCC (COP15), held in Copenhagen in 2009, the Copenhagen Accord was agreed in order to commit to the goal of sending $100 billion per year to developing countries in assistance for climate change mitigation and adaptation through 2020. A new fund - the Green Climate Fund, was therefore created.

Additionality

A key and defining feature of international adaptation finance is its premise on the concept of additionality. This reflects the linkages between adaptation finance and other levels of development aid. Many developing countries already provide international aid assistance to developing countries to address challenges such as poverty, malnutrition, food insecurity, availability of drinking water, indebtedness, illiteracy, unemployment, local resource conflicts, and lower technological development. Climate change threatens to exacerbate or stall progress on fixing some of these pre-existing problems, and creates new problems. To avoid existing aid being redirected, additionality refers to the extra costs of adaptation.

The four main definitions of additionality are:

- Climate finance classified as aid, but additional to (over and above) the 0.7% ODA target;

- Increase on previous year's Official Development Assistance (ODA) spent on climate change mitigation;

- Rising ODA levels that include climate change finance but where it is limited to a specified percentage; and

- Increase in climate finance not connected to ODA.

A criticism of additionality is that it encourages business as usual that does not account for the future risks of climate change. Some advocates have thus proposed integrating climate change adaptation into poverty reduction programs.

Other Adaptation Finance Mechanisms

There are several other climate change adaptation finance proposals, most of which employ official development assistance or ODA. These proposals range from a World Bank program, to proposals involving auctioning of carbon allowances, to a global carbon or transportation tax, to compensation-based funding. Other proposals suggest using market-based mechanisms, rather than ODA, such as the Higher Ground Foundation's vulnerability reduction credit (VRC) or a program similar to the Clean Development Mechanism, to raise private money for climate change adaptation.

Considerations and General Recommendations

Principles for Effective Policy

Adaptive policy can occur at the global, national, or local scale, with outcomes dependent on the political will in that area. Scheraga and Grambsch identify 9 fundamental principles to be considered when designing adaptation policy:

- The effects of climate change vary by region.

- The effects of climate change may vary across demographic groups.

- Climate change poses both risks and opportunities.

- The effects of climate change must be considered in the context of multiple stressors and factors, which may be as important to the design of adaptive responses as the sensitivity of the change.

- Adaptation comes at a cost.

- Adaptive responses vary in effectiveness, as demonstrated by current efforts to cope with climate variability.

- The systemic nature of climate impacts complicates the development of adaptation policy.

- Maladaptation can result in negative effects that are as serious as the climate-induced effects that are being avoided.

- Many opportunities for adaptation make sense whether or not the effects of climate change are realized.

Scheraga and Grambsch make it clear that climate change policy is impeded by the high level of variance surrounding climate change impacts as well as the diverse nature of the problems they face.

Adaptation can mitigate the adverse impacts of climate change, but it will incur costs and will not prevent all damage. The IPCC points out that many adverse effects of climate change are not changes in the average conditions, but changes in the variation or the extremes of conditions. For example, the average sea level in a port might not be as important as the height of water during a storm surge (which causes flooding); the average rainfall in an area might not be as important as how frequent and severe droughts and extreme precipitation events become. Additionally, effective adaptive policy can be difficult to implement because policymakers are rewarded more for enacting short-term change, rather than long-term planning. Since the impacts of climate change are generally not seen in the short-term, this means that policymakers have less incentive to act upon those potential outcomes. Furthermore, these problems (both the causes and effects of climate change) are occurring on a global scale, which has caused the United Nations to lead global policy efforts such as the Kyoto Protocol and Paris Agreement, in addition to creating a body of research through the IPCC, in order to create a global framework for adapting to and combatting climate change. However, the vast majority of climate change adaptation and mitigation policies are being implemented on a more local scale due to the fact that different regions must adapt differently to climate change and because national and global policies are often more challenging to enact.

Criteria for Assessing Responses

James Titus, project manager for sea level rise at the U.S. Environmental Protection Agency, identifies the following criteria that policy makers should use in assessing responses to global warming:

- Economic Efficiency: Will the initiative yield benefits substantially greater than if the resources were applied elsewhere?

- Flexibility: Is the strategy reasonable for the entire range of possible changes in temperatures, precipitation, and sea level?

- Urgency: Would the strategy be successful if implementation were delayed ten or twenty years?

- Low Cost: Does the strategy require minimal resources?

- Equity: Does the strategy unfairly benefit some at the expense of other regions, generations, or economic classes?

- Institutional feasibility: Is the strategy acceptable to the public? Can it be implemented with existing institutions under existing laws?

- Unique or Critical Resources: Would the strategy decrease the risk of losing unique environmental or cultural resources?

- Health and Safety: Would the proposed strategy increase or decrease the risk of disease or injury?

- Consistency: Does the policy support other national state, community, or private goals?

- Private v. Public Sector: Does the strategy minimize governmental interference with decisions best made by the private sector?

Differing Time Scales

Adaptation can either occur in anticipation of change (anticipatory adaptation), or be a response to those changes (reactive adaptation). Most adaptation being implemented at present is responding to current climate trends and variability, for example increased use of artificial snow-making in the European Alps. Some adaptation measures, however, are anticipating future climate change, such as the construction of the Confederation Bridge in Canada at a higher elevation to take into account the effect of future sea-level rise on ship clearance under the bridge.

Much adaptation takes place in relation to short-term climate variability, however this may cause maladaptation to longer-term climatic trends. For example, the expansion of irrigation in Egypt into the Western Sinai desert due to a period of higher river flows is a maladaptation when viewed in relation to the longer term projections of drying in the region). Adaptations at one scale can also create externalities at another by reducing the adaptive capacity of other actors. This is often the case when broad assessments of the costs and benefits of adaptation are examined at smaller scales and it is possible to see that whilst the adaptation may benefit some actors, it has a negative effect on others.

Traditional Coping Strategies

People have always adapted to climatic changes and some community coping strategies already exist, for example changing sowing times or adopting new water-saving techniques. Traditional knowledge and coping strategies must be maintained and strengthened, otherwise adaptive capacity may be weakened as local knowledge of the environment is lost. Strengthening these local techniques and building upon them also makes it more likely that adaptation strategies will be adopted, as it creates more community ownership and involvement in the process. In many cases however this will not be enough to adapt to new conditions which are outside the range of those previously experienced, and new techniques will be needed. The incremental adaptations which were being implemented are now insufficient as the vulnerabilities and risks of climate change have increased, this causes a need for transformational adaptations which are much larger and costlier. Current development efforts are increasingly focusing on community-based climate change adaptation, seeking to enhance local knowledge, participation and ownership of adaptation strategies.

Types of Adaptation

Local Adaptation Efforts

Cities, states, and provinces often have considerable responsibility in land use planning, public health, and disaster management. Some have begun to take steps to adapt to threats intensified by climate change, such as flooding, bushfires, heatwaves, and rising sea levels.

Projects include:

- Installing protective and resilient technologies and materials in properties that are prone to flooding.

- Changing to heat tolerant tree varieties (Chicago).

- Rainwater storage to deal with more frequent flooding rainfall – Changing to water permeable pavements, adding water-buffering vegetation, adding underground storage tanks, subsidizing household rain barrels (Chicago).

- Reducing paved areas to deal with rainwater and heat (Chicago, Seoul).

- Adding green roofs to deal with rainwater and heat (Chicago).

- Adding air conditioning in public schools (Chicago).

- Requiring waterfront properties to have higher foundations (Chula Vista, California).

- Raising pumps at wastewater treatment plants (New York City).

- Surveying local vulnerabilities, raising public awareness, and making climate change-specific planning tools like future flood maps (Seattle, Chicago, Norfolk, many others).

- Incentivizing lighter-colored roofs to reduce the heat island effect (Chula Vista, California).

- Installing devices to prevent seawater from backflowing into storm drains (San Francisco).

- Installing better flood defenses, such as sea walls and increased pumping capacity (Miami Beach).

- Buying out homeowners in flood-prone areas (New Jersey).

- Raising street level to prevent flooding (Miami Beach).

Dealing with more frequent drenching rains may required increasing the capacity of stormwater systems, and separating stormwater from blackwater, so that overflows in peak periods do not contaminate rivers. One example is the SMART Tunnel in Kuala Lumpur.

According to English Nature, gardeners can help mitigate the effects of climate change by providing habitats for the most threatened species, and saving water by changing gardens to use plants which require less.

New York City produced a comprehensive report for its Rebuilding and Resiliency initiative after Hurricane Sandy. Its efforts include not only making buildings less prone to flooding, but taking steps to reduce the future recurrence of specific problems encountered during and after the storm: weeks-long fuel shortages even in unaffected areas due to legal and transportation problems, flooded health care facilities, insurance premium increases, damage to electricity and steam generation in addition to distribution networks, and flooding of subway and roadway tunnels.

Enhancing Adaptive Capacity

Adaptive capacity is the ability of a system (human, natural or managed) to adjust to climate change (including climate variability and extremes) to moderate potential damages, to take advantage of opportunities, or to cope with consequences. As a property, adaptive capacity is distinct from adaptation itself. Those societies that can respond to change quickly and successfully have a high adaptive capacity. High adaptive capacity does not necessarily translate into successful adaptation. For example, adaptive capacity in Western Europe is generally considered to be high, and the risks of warmer winters increasing the range of livestock diseases is well documented, but many parts of Europe are still badly affected by outbreaks of the Bluetongue virus in livestock in 2007.

Unmitigated climate change (i.e., future climate change without efforts to limit greenhouse gas emissions) would, in the long term, be likely to exceed the capacity of natural, managed and human systems to adapt.

It has been found that efforts to enhance adaptive capacity can help to reduce vulnerability to climate change. In many instances, activities to promote sustainable development can also act to enhance people's adaptive capacity to climate change. These activities can include:

- Improving access to resources,
- Reducing poverty,
- Lowering inequities of resources and wealth among groups,
- Improving education and information,
- Improving infrastructure,
- Improving institutional capacity and efficiency,
- Promoting local indigenous practices, knowledge, and experiences.

Others have suggested that certain forms of gender inequity should be addressed at the same time; for example women may have participation in decision-making, or be constrained by lower levels of education.

Researchers at the Overseas Development Institute found that development interventions to increase adaptive capacity have tended not to result in increased agency for local people. They argue that this should play a more prominent part in future intervention planning because agency is a central factor in all other aspects of adaptive capacity. Asset holdings and the ability to convert these resources through institutional and market processes are central to agency.

Agricultural Production

A significant effect of global climate change is the altering of global rainfall patterns, with certain effects on agriculture. Rainfed agriculture constitutes 80% of global agriculture. Many of the 852 million poor people in the world live in parts of Asia and Africa that depend on rainfall to cultivate food crops. Climate change will modify rainfall, evaporation, runoff, and soil moisture storage. Extended drought can cause the failure of small and marginal farms with resultant economic, political and social disruption, more so than this currently occurs.

Agriculture of any kind is strongly influenced by the availability of water. Changes in total seasonal precipitation or in its pattern of variability are both important. The occurrence of moisture stress during flowering, pollination, and grain-filling is harmful to most crops and particularly so to corn, soybeans, and wheat. Increased evaporation from the soil and accelerated transpiration in the plants themselves will cause moisture stress.

Adaptive ideas include:

- Taking advantage of global transportation systems to delivering surplus food to where it is needed (though this does not help subsistence farmers unless aid is given).

- Developing crop varieties with greater drought tolerance.

- Rainwater storage. For example, according to the International Water Management Institute, using small planting basins to 'harvest' water in Zimbabwe has been shown to boost maize yields, whether rainfall is abundant or scarce. And in Niger, they have led to three or fourfold increases in millet yields.

- Falling back from crops to wild edible fruits, roots and leaves. Promoting the growth of forests can provide these backup food supplies, and also provide watershed conservation, carbon sequestration, and aesthetic value.

Reforestation

Reforestation is one of the ways to stop desertification fueled by anthropogenic climate change and non sustainable land use. One of the most important projects is the Great Green Wall that should stop the expansion of Sahara desert to the south. By 2018 only 15% of it is accomplished, but there are already many positive effects, which include: "Over 12 million acres (5 million hectares) of degraded land has been restored in Nigeria; roughly 30 million acres of drought-resistant trees have been planted across Senegal; and a whopping 37 million acres of land has been restored in Ethiopia – just to name a few of the states involved." Many groundwater wells refilled with drinking water, rural towns with additional food supplies, and new sources of work and income for villagers, thanks to the need for tree maintenance.

More Spending on Irrigation

The demand for water for irrigation is projected to rise in a warmer climate, bringing increased competition between agriculture—already the largest consumer of water resources in semi-arid regions—and urban as well as industrial users. Falling water tables and the resulting increase in the energy needed to pump water will make the practice of irrigation more expensive, particularly when with drier conditions more water will be required per acre. Other strategies will be needed to make the most efficient use of water resources. For example, the International Water Management Institute has suggested five strategies that could help Asia feed its growing population in light of climate change. These are:

- Modernising existing irrigation schemes to suit modern methods of farming,

- Supporting farmers' efforts to find their own water supplies, by tapping into groundwater in a sustainable way,

- Looking beyond conventional "Participatory Irrigation Management" schemes, by engaging the private sector,

- Expanding capacity and knowledge,

- Investing outside the irrigation sector.

Weather Control

Russian and American scientists have in the past tried to control the weather, for example by seeding clouds with chemicals to try to produce rain when and where it is needed. A new method being developed involves replicating the urban heat island effect, where cities are slightly hotter than the countryside because they are darker and absorb more heat. This creates 28% more rain 20–40 miles downwind from cities compared to upwind. On the timescale of several decades, new weather control techniques may become feasible which would allow control of extreme weather such as hurricanes.

The World Meteorological Organization (WMO) through its Commission for Atmospheric Sciences (CAS) opined in 2007: "Purposeful augmentation of precipitation, reduction of hail damage, dispersion of fog and other types of cloud and storm modifications by cloud seeding are developing technologies which are still striving to achieve a sound scientific foundation and which have to be adapted to enormously varied natural conditions."

Damming Glacial Lakes

Glacial lake outburst floods may become a bigger concern due to the retreat of glaciers, leaving behind numerous lakes that are impounded by often weak terminal moraine dams. In the past, the sudden failure of these dams has resulted in localized property damage, injury and deaths. Glacial lakes in danger of bursting can have their moraines replaced with concrete dams (which may also provide hydroelectric power).

Geoengineering

IPCC concluded that geoengineering options, such as ocean fertilization to remove CO_2 from the atmosphere, remained largely unproven. It was judged that reliable cost estimates for geoengineering had not been published.

The Royal Society published the findings of a study into geoengineering. The authors of the study defined geoengineering as a "deliberate large-scale intervention in the Earth's climate system, in order to moderate global warming". According to the study, the safest and most predictable method of moderating climate change is early action to reduce GHG emissions.

Scientists such as Ken Caldeira and Paul Crutzen suggest techniques such as:

- Solar radiation management may be seen as an adaptation to global warming. Techniques such as space sunshade, creating stratospheric sulfur aerosols and painting roofing and paving materials white all fall into this category.

- Hydrological geoengineering - typically seeking to preserve sea ice or adjust thermohaline circulation by using methods such as diverting rivers to keep warm water away from sea ice, or tethering icebergs to prevent them drifting into warmer waters and melting. Though this is an adaptation technique, if it prevents Arctic methane release it would also be classified as mitigation.

Migration

Migration frequently requires would-be migrants to have access to social and financial capital, such as support networks in the chosen destination, and the funds or physical resources to be able to move. It is frequently the last adaptive response households will take when confronted with environmental factors that threaten their livelihoods, and mostly resorted to when other mechanisms to cope have proven unsuccessful.

The rhetoric of migration being related to climate change is complex and disputed. However, it is widely accepted that the results of migration events are multi-causal, with the environment being just a factor amongst many. Outside of policy, human rights organizations, expert demographers and environmental climate scientists dominate this debate. Many discussions are based on projections and less with current migration data. While many migration events can be attributed to sudden environmental change, most migration events are a result of long term environmental changes and do not cause sudden migration. Some scholars attribute these events to sudden environmental changes, like natural disasters. Some choose to label it "climate change", which reflects a more long term onset of change, and the human impact element.

It is helpful to provide an intersectional approach to this discussion and understand that focusing on climate change as the issue frames the debate in terms of projections, causing the research to be speculative. Migration as tool for climate change adaptation is projected to be a more pressing issue in the decade to come. It is often framed in terms of human rights issues and national security. Migration events are often seen as a failure of the governments or policy making bodies that could not contain or effectively manage environmental changes. For example, extreme drought events in the Caribbean proliferate movement of peoples because of the lack of water. This is often seen as a failure on the local governments to provide structural and independent resources. These adaptation failures that have been the topic of concern for many scholars researching this area. The UN High Commissioner for Refugees has been viewed as one of the highest authorities to help those displaced. In Africa, specifically, migrant social networks can help to build social capital to increase the social resilience in the communities of origin and trigger innovations across regions by the transfer of knowledge, technology, remittances and other resources. These could increase the flexibility, diversity and creativity of communities in addressing climate stress and open new pathways for co-development connecting the home and host communities.

In Africa, in particular, in terms of adaptation strategies Mozambique and Zimbabwe are clear examples of this because they have implemented relocation policies that have reduced the exposure of populations and migrants to disaster. In any case, it is important to build resilience in the long run. And for that, tools must be put in place that limit forced displacement after a disaster; promote employment programs, even if only temporary, for IDPs or establish funding plans to ensure their security; to minimize the vulnerability of populations from risk areas. This can limit the displacement caused by environmental shocks and better channel the positive spillovers (money transfers, experiences, etc.) from the migration to the origin countries/communities.

The figure of the "failed migrant", in most African countries, shows extreme heterogeneity. The causes associated with failure are most often from social and personal natures – feelings of personal failure for example– but can also be related to social isolation in the host countries. Although there has been some progress in the discussion of the causes of the pathos of failed migration, there are still many unresolved issues. Factors such as a low social level, a change of life plan, unemployment, or even environmental stress (drought, high temperature, water scarcity, etc.) are often associated with an increased risk of failure when we know that most African migrants live in difficult socio-economic and ecological conditions.

Insurance

Insurance spreads the financial impact of flooding and other extreme weather events. Although it can be preferable to take a proactive approach to eliminate the cause of the risk, reactive post-harm compensation can be used as a last resort. Access to reinsurance may be a form of increasing the resiliency of cities. Where there are failures in the private insurance market, the public sector can subsidize premiums. A study identified key equity issues for policy considerations:

- Transferring risk to the public purse does not reduce overall risk,

- Governments can spread the cost of losses across time rather than space,

- Governments can force home-owners in low risk areas to cross-subsidize the insurance premiums of those in high risk areas,

- Cross-subsidization is increasingly difficult for private sector insurers operating in a competitive market,

- Governments can tax people to pay for tomorrow's disaster.

Government-subsidized insurance, such as the U.S. National Flood Insurance Program, is criticized for providing a perverse incentive to develop properties in hazardous areas, thereby increasing overall risk. It is also suggested that insurance can undermine other efforts to increase adaptation, for instance through property level protection and resilience. This behavioral effect may be countered with appropriate land-use policies that limit new construction where current or future climate risks are perceived and/or encourage the adoption of resilient building codes to mitigate potential damages.

Complementary to Mitigation

IPCC Working Group II, the United States National Academy of Sciences, the United Nations Disaster Risk Reduction Office, and other science policy experts agree that while mitigating the emission of greenhouse gases is important, adaptation to the effects of global warming will still be necessary. Some, like the UK Institution of Mechanical Engineers, worry that mitigation efforts will largely fail. The IPCC group points out that the world's ability to mitigate global warming is an economic and political challenge. Given that greenhouse gas levels are already elevated, the lag of decades between emissions and some impacts, and the significant economic and political challenges of success, the IPCC group points out that it is uncertain how much climate change will be mitigated.

Developing countries are the least able to adapt to climate change. Doing so depends on such factors as wealth, technology, education, infrastructure, access to resources, management capabilities, acceptance of the existence of climate change and the consequent need for action, and sociopolitical will.

After assessing the literature on sustainability and climate change scientists concluded with high confidence that up to the year 2050, an effort to cap GHG emissions at 550 ppm would benefit developing countries significantly. This was judged to be especially the case when combined with enhanced adaptation. By 2100, however, it was still judged likely that there would be significant climate change impacts. This was judged to be the case even with aggressive mitigation and significantly enhanced adaptive capacity.

The IPCC group also pointed out that climate change adaptation measures can reinforce and be reinforced by efforts to promote sustainable development and reduce poverty.

Cost of Adaptation vs. Mitigation

Adaptation and mitigation can be viewed as two competing policy responses, with tradeoffs between the two. The other tradeoff is with climate change impacts. In practice, however, the actual tradeoffs are debatable. This is because the people who bear emission reduction costs or benefits are often different from those who pay or benefit from adaptation measures.

Economists, using cost-benefit analysis, have attempted to calculate an "optimal" balance of the costs and benefits between climate change impacts, adaptation, and mitigation. There are difficulties in doing this calculation, for example, future climate change damages are uncertain, as are the future costs of adaptation.

Also, deciding what "optimal" is depends on value judgements made by the economist doing the study. For example, how to value impacts occurring in different regions and different times, and "non-market" impacts, e.g., damages to ecosystems. Economics cannot provide definitive answers to these questions over valuation, and some valuations may be viewed as being controversial.

Some reviews indicate that policymakers are uncomfortable with using the results of this type of economic analysis. This is due to the uncertainties surrounding cost estimates for climate change damages, adaptation, and mitigation. Another type of analysis is based on a risk-based approach to the problem. It has been argued that adaptation could play an important role in climate policy, but not in an explicit trade-off against mitigation.

Various attempts have been made to estimate the cost of adaptation to climate change. An early study by the UNFCCC in 2007 estimated costs of adaptation to be $49–171 billion per year globally by 2030, of which a significant share of the additional investment and financial flows, USD $28–67 billion would be needed in 2030 in non-Annex I Parties. This represents a doubling of current official development assistance (ODA). However, this estimate was criticised as underestimating costs of adaptation by a factor of 2 or 3, particularly as it did not take into account sectors such as tourism, mining, energy and retail. More recent estimates by the United Nations Development Programme and World Bank put estimates between $75-100 billion per year between 2010 and 2050, depending on the scenario of climate change.

The benefits of strong, early action on mitigation considerably outweigh the costs.

Opposition to Adaptation

According to Al Gore, writing in 1992 in *Earth in the Balance*, adaptation represented a "kind of laziness, an arrogant faith in our ability to react in time to save our skins". Climate commentator David Roberts has written that "both mitigation and adaption are necessary at this point. But for every day mitigation is delayed, the need for adaptation grows," which is problematic because "adaptation is more expensive and requires bigger government than mitigation."

The are some measure that seems like adaptation, but can lead, in fact, to more climate sensitivity and more climate change. For example, reliance on air conditioning gives relief from the heat, but can create an addiction to it, e. g. intolerance to any uncomfortable climate. Air conditioning also consumes 20% of all energy in United States therefore emitting a lot of greenhouse gases into the atmosphere, according to the book *Losing Our Cool* of Stan Cox.

Climate Adaptation Denial

According to a report released by Greenpeace USA in September 2013, climate change denial and the campaigns designed to block adaptation measures grew mainly out of the 1990s negotiations slated to develop a global agreement. During these talks, a number of lobby groups were established with an objective of developing doubt within policymakers and the media through the use of publications in the guise of true science. This tactic, similar to those of large tobacco companies, was utilized by the lobby groups in the hopes of delaying action and blurring the lines between the valid scientific efforts to challenge climate change findings and those designed to merely undermine the credibility of the scientific community. This strategy feeds into the "uncertainty argument" and develops an impression of debate through references to the uncertainty of scientific findings that exist in any research model. Additional tactics that the lobbyist groups have used include releasing non-stories manufactured from stolen emails and communications plans to develop more media coverage of the uncertainty argument.

Effects of Climate Change

Climate change has evident effects on ecosystems and people. Here's a list of the climate change phenomena we're already experiencing first-hand.

Ice Melting

One of the most evident consequences is melting ice – the melting of the cryosphere, those portions of Earth's surface where water is in solid form, including ice caps, glaciers, and permafrost (those areas where soil is permanently frozen). According to predictions, Arctic ice could completely melt during the hottest periods of the year by the end of the century. The cryosphere naturally plays a crucial role in the global climate system and a change in its extension could cause a change in the system itself. Fragile ecosystems like oceans, mountains and wetlands could be damaged permanently.

Sea Level Rise

Melting ice caps in Antarctica and Greenland have most likely led sea level to rise by 3.1 millimetres per year between 1993 and 2003, according to IPCC. The rise is expected to reach 15 to 95 centimetres by 2100.

Ocean Acidification

Increased levels of CO_2 in the atmosphere will also lead to ocean acidification, causing irreparable damage to marine ecosystems – like the Great Barrier Reef that is a UNESCO heritage site as it is home to more than "400 types of coral, 1,500 species of fish and 4,000 types of mollusc. It also holds great scientific interest as the habitat of species such as the dugong and the large green turtle, which are threatened with extinction". Professional services firm Deloitte estimated the economic value of this treasure: 56 billion Australian dollars, and 64,000 jobs.

Desertification

Desertification (and, thus, heat waves) will expand to areas that currently boast a temperate climate such as the areas north and south of the Sahara desert, including the Mediterranean countries, causing severe damage to agriculture. Crops will significantly drop while more and more people will face undernourishment. In particular, yields from maize and wheat crops could drop by 50 per cent over the next 35 years due to global warming. It's a risk that has to be prevented considering that people suffering from hunger are currently slightly decreasing. The study State of Food Insecurity in the World 2015 conducted by the International Fund for Agricultural development (IFAD) and the World Food Programme (WFP) estimates that nearly 795 million people don't eat enough food. They amounted to 1 billion in 1990-1992.

Desertification will axpand to areas now boasting a temperate climate.

Events like El Niño – a variation in the southern oscillation that causes significant changes in climate including hurricanes, storms, flooding in Central America and severe drought linked to wild fires in western Pacific areas – will be more frequent and intense causing casualties and economic loss. This could lead to the outburst of diseases, like malaria, in areas previously unaffected.

Biodiversity Loss

It's not only due to climate change, but also because of humanity, that the Earth is facing a relentless mass extinction, the sixth, resulting in a significant drop in our Planet's biodiversity. Species extinction rate is extremely high and half the living species could become extinct by the end of the century. This biodiversity loss has grave and far-reaching implications for human well-being. As it happens with desertification, biodiversity loss – in particular of plant species – could slow down disease control and increase the spreading of infectious and autoimmune diseases.

Climate Impacts on Human Health

The impacts of climate change include warming temperatures, changes in precipitation, increases in the frequency or intensity of some extreme weather events, and rising sea levels. These impacts threaten our health by affecting the food we eat, the water we drink, the air we breathe, and the weather we experience.

The severity of these health risks will depend on the ability of public health and safety systems to address or prepare for these changing threats, as well as factors such as an individual's behavior, age, gender, and economic status. Impacts will vary based on a where a person lives, how sensitive they are to health threats, how much they are exposed to climate change impacts, and how well they and their community are able to adapt to change.

People in developing countries may be the most vulnerable to health risks globally, but climate change poses significant threats to health even in wealthy nations.

Temperature-related Impacts

Warmer average temperatures will lead to hotter days and more frequent and longer heat waves. These changes will lead to an increase in heat-related deaths in the United States—reaching as much as thousands to tens of thousands of additional deaths each year by the end of the century during summer months.

These deaths will not be offset by the smaller reduction in cold-related deaths projected in the winter months. However, adaptive responses, such as wider use of air conditioning, are expected to reduce the projected increases in death from extreme heat.

Exposure to extreme heat can lead to heat stroke and dehydration, as well as cardiovascular, respiratory, and cerebrovascular disease. Excessive heat is more likely to affect populations in northern latitudes where people are less prepared to cope with excessive temperatures. Certain types of populations are more vulnerable than others: for example, outdoor workers, student athletes, and homeless people tend to be more exposed to extreme heat because they spend more time outdoors. Low-income households and older adults may lack access to air conditioning which also increases exposure to extreme heat. Additionally, young children, pregnant women, older adults, and people with certain medical conditions are less able to regulate their body temperature and can therefore be more vulnerable to extreme heat.

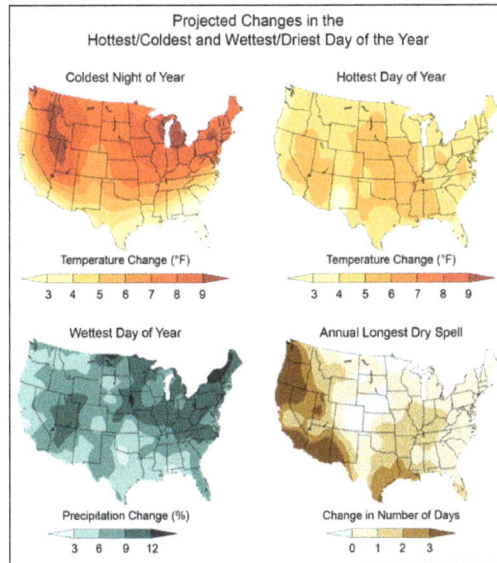

Projected Changes in the
Hottest/Coldest and Wettest/Driest Day of the Year

Projected changes in several climate variables for 2046-2065 with respect to the 1981-2000 average for the RCP6.0 scenario. These include the coldest night of the year (top left) and the hottest day of the year (top right). By the middle of this century, the coldest night of the year is projected to warm by 6°F to 10°F over most of the country, with slightly smaller changes in the south. The warmest day of the year is projected to be 4°F to 6°F warmer in most areas. Also shown are projections of the wettest day of the year (bottom left) and the annual longest consecutive dry day spell (bottom right). Extreme precipitation is projected to increase, with an average change of 5% to 15% in the precipitation falling on the wettest day of the year. The length of the annual longest dry spell is projected to increase in most areas, but these changes are small: less than two days in most areas.

Urban areas are typically warmer than their rural surroundings. Large metropolitan areas such as St. Louis, Philadelphia, Chicago, and Cincinnati have seen notable increases in death rates during heat waves. Climate change is projected to increase the vulnerability of urban populations to heat-related health impacts in the future. Heat waves are also often accompanied by periods of stagnant air, leading to increases in air pollution and associated health effects.

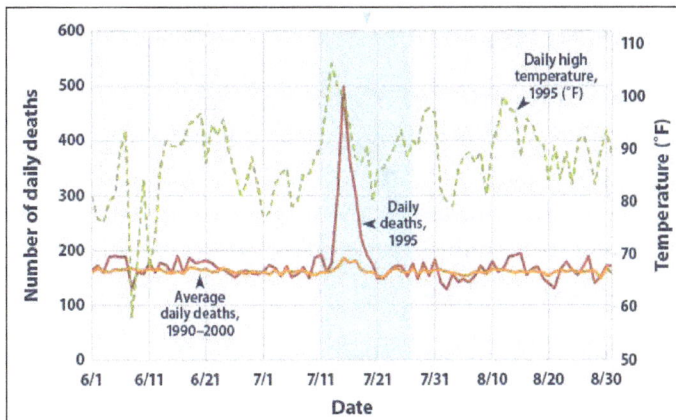

This figure shows the relationship between high temperatures and deaths observed during the 1995 Chicago heat wave. The large spike in deaths in mid-July (red line) is much higher than the average number of deaths during that time of year (orange line), as well as the death rate before and after the heat wave.

Air Quality Impacts

Changes in the climate affect the air we breathe both indoors and outdoors. Warmer temperatures and shifting weather patterns can worsen air quality, which can lead to asthma attacks and other respiratory and cardiovascular health effects. Wildfires, which are expected to continue to increase in number and severity as the climate changes, create smoke and other unhealthy air pollutants. Rising carbon dioxide levels and warmer temperatures also affect airborne allergens, such as ragweed pollen.

Despite significant improvements in U.S. air quality since the 1970s, as of 2014 about 57 million Americans lived in counties that did not meet national air quality standards. Climate change may make it even harder for states to meet these standards in the future, exposing more people to unhealthy air.

Increases in Ozone

Scientists project that warmer temperatures from climate change will increase the frequency of days with unhealthy levels of ground-level ozone, a harmful air pollutant, and a component in smog.

- People exposed to higher levels of ground-level ozone are at greater risk of dying prematurely or being admitted to the hospital for respiratory problems.

- Ground-level ozone can damage lung tissue, reduce lung function, and inflame airways. This can aggravate asthma or other lung diseases. Children, older adults, outdoor workers, and those with asthma and other chronic lung diseases are particularly at risk.

- Because warm, stagnant air tends to increase the formation of ozone, climate change is likely to increase levels of ground-level ozone in already-polluted areas of the United States and increase the number of days with poor air quality.

- The higher concentrations of ozone due to climate change may result in tens to thousands of additional ozone-related illnesses and premature deaths per year by 2030 in the United States, assuming no change in projected air quality policies.

Changes in Particulate Matter

Particulate matter is the term for a category of extremely small particles and liquid droplets suspended in the atmosphere. Fine particles include those smaller than 2.5 micrometers (about one ten-thousandth of an inch). Some particulate matter such as dust, wildfire smoke, and sea spray occur naturally, while some is created by human activities such as the burning of fossil fuels to produce energy. These particles may be emitted directly or may be formed in the atmosphere from chemical reactions of gases such as sulfur dioxide, nitrogen dioxide, and volatile organic compounds.

- Inhaling fine particles can lead to a broad range of adverse health effects, including lung cancer, chronic obstructive pulmonary disease (COPD), and cardiovascular disease.

- Climate change is expected to increase the number and severity of wildfires. Particulate matter from wildfire smoke can often be carried very long distances by the wind, affecting people who live far from the source of this air pollutant.

- Older adults are particularly sensitive to short-term particle exposure, with a higher risk of hospitalization and death. Outdoor workers like firefighters can also have high exposure.

Due to the complex factors that influence atmospheric levels of fine particulate matter, scientists do not yet know whether climate change will increase or decrease particulate matter concentrations across the United States. Particulate matter can be removed from the air by rainfall, and precipitation is expected to increase in quantity though not necessarily frequency. Climate-related changes in stagnant air episodes, wind patterns, emissions from vegetation and the chemistry of atmospheric pollutants will also affect particulate matter levels.

Changes in Allergens and Asthma Triggers

Allergic illnesses, including hay fever, affect about one-third of the U.S. population, and more than 34 million Americans have been diagnosed with asthma. Climate change may affect allergies and respiratory health. The spring pollen season is already occurring earlier in the United States for certain types of plants, and the length of the season has increased for some plants with highly allergenic pollen such as ragweed. In addition to lengthening the ragweed pollen season, rising carbon dioxide concentrations and temperatures may also lead to earlier flowering, more flowers, and increased pollen levels in ragweed.

Impacts from Extreme Weather Events

Hurricane Katrina was one of the most devastating hurricanes in the United States, responsible for an estimated 971 to 1,300 deaths.

Increases in the frequency or severity of some extreme weather events, such as extreme precipitation, flooding, droughts, and storms, threaten the health of people during and after the event. The people most at risk include young children, older adults, people with disabilities or medical conditions, and the poor. Extreme events can affect human health in a number of ways by:

- Reducing the availability of safe food and drinking water.

- Damaging roads and bridges, disrupting access to hospitals and pharmacies.

- Interrupting communication, utility, and health care services.

- Contributing to carbon monoxide poisoning from improper use of portable electric generators during and after storms.

- Increasing stomach and intestinal illness, particularly following power outages.

- Creating or worsening mental health impacts such as depression and post-traumatic stress disorder (PTSD).

In addition, emergency evacuations pose health risks to older adults, especially those with limited mobility who cannot use elevators during power outages. Evacuations may be complicated by the need for concurrent transfer of medical records, medications, and medical equipment. Some individuals with disabilities may also be disproportionally affected if they are unable to access evacuation routes, have difficulty in understanding or receiving warnings of impending danger, or have limited ability to communicate their needs.

Vectorborne Diseases

Vectorborne diseases are illnesses that are transmitted by disease *vectors*, which include mosquitoes, ticks, and fleas. These vectors can carry infectious pathogens, such as viruses, bacteria, and protozoa, from animals to humans. Changes in temperature, precipitation, and extreme events increases the geographic range of diseases spread by vectors and can lead to illnesses occurring earlier in the year.

- The geographic range of ticks that carry Lyme disease is limited by temperature. As air temperatures rise, ticks are likely to become active earlier in the season, and their range is likely to continue to expand northward. Typical symptoms of Lyme disease include fever, headache, fatigue, and a characteristic skin rash.

- Mosquitoes thrive in certain climate conditions and can spread diseases like West Nile virus. Extreme temperatures—too cold, hot, wet, or dry—influence the location and number of mosquitoes that transmit West Nile virus. More than three million people were estimated to be infected with West Nile virus in the United States from 1999 to 2010.

The spread of climate-sensitive diseases will depend on both climate and non-climate factors such as land use, socioeconomic and cultural conditions, pest control, access to health care, and human responses to disease risk. The United States has public health infrastructure and programs to monitor, manage, and prevent the spread of many diseases. The risks for climate-sensitive diseases can be much higher in poorer countries that have less capacity to prevent and treat illness.

West Nile virus is maintained in transmission cycles between birds (the natural hosts of the virus) and mosquitoes. Human infections can occur from a bite of a mosquito that has previously bitten an infected bird. Warmer winters, longer frost-free season, and earlier spring arrival may influence the migration patterns and fledgling survival of birds that are the natural host of West Nile virus. In addition, rising temperature, changing precipitation patterns, and a higher frequency of extreme weather events are likely to influence the distribution and abundance of mosquitoes that transmit West Nile virus.

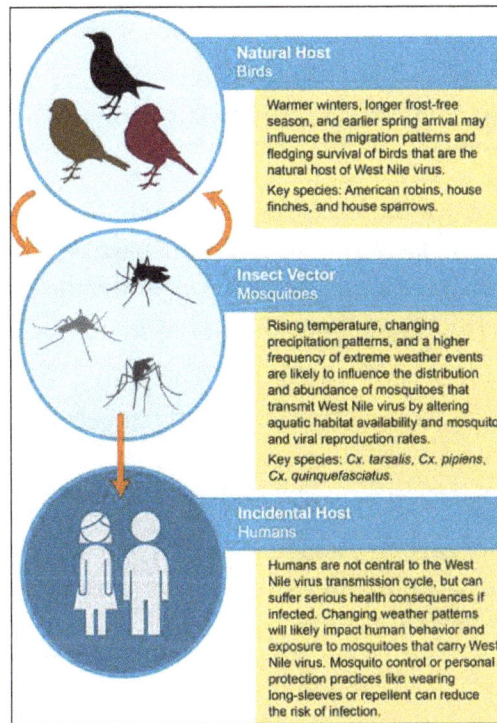

Water-related Illnesses

People can become ill if exposed to contaminated drinking or recreational water. Climate change increases the risk of illness through increasing temperature, more frequent heavy rains and runoff, and the effects of storms. Health impacts may include gastrointestinal illness like diarrhea, effects on the body's nervous and respiratory systems, or liver and kidney damage.

- Climate impacts can affect exposure to waterborne pathogens (bacteria, viruses, and parasites such as *Cryptosporidium* and *Giardia*); toxins produced by harmful algal and cyanobacterial blooms in the water; and chemicals that end up in water from human activities.

- Changing water temperatures mean that waterborne *Vibrio* bacteria and harmful algal toxins will be present in the water or in seafood at different times of the year, or in places where they were not previously threats.

- Runoff and flooding resulting from increases in extreme precipitation, hurricane rainfall, and storm surge will increasingly contaminate water bodies used for recreation (such as lakes and beaches), shellfish harvesting waters, and sources of drinking water.

- Extreme weather events and storm surges can damage or exceed the capacity of water infrastructure (such as drinking water or wastewater treatment plants), increasing the risk that people will be exposed to contaminants.

Water resource, public health, and environmental agencies in the United States provide many public health safeguards to reduce risk of exposure and illness even if water becomes contaminated. These include water quality monitoring, drinking water treatment standards and practices, beach closures, and issuing advisories for boiling drinking water and harvesting shellfish.

Food Safety and Nutrition

Climate change and the direct impacts of higher concentrations of carbon dioxide in the atmosphere are expected to affect food safety and nutrition. Extreme weather events can also disrupt or slow the distribution of food.

- Higher air temperatures can increase cases of *Salmonella* and other bacteria-related food poisoning because bacteria grow more rapidly in warm environments. These diseases can cause gastrointestinal distress and, in severe cases, death. Practices to safeguard food can help avoid these illnesses even as the climate changes.

- Climate change will have a variety of impacts that may increase the risk of exposure to chemical contaminants in food. For example, higher sea surface temperatures will lead to higher mercury concentrations in seafood, and increases in extreme weather events will introduce contaminants into the food chain through stormwater runoff.

- Higher concentrations of carbon dioxide in the air can act as a "fertilizer" for some plants, but lowers the levels of protein and essential minerals in crops such as wheat, rice, and potatoes, making these foods less nutritious.

- Extreme events, such as flooding and drought, create challenges for food distribution if roads and waterways are damaged or made inaccessible.

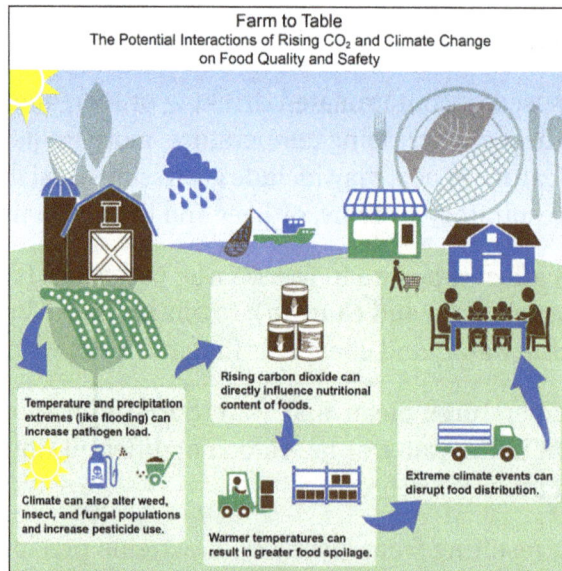

The food system involves a network of interactions with our physical and biological environments as food moves from production to consumption, or from "farm to table." Rising CO_2 and climate change will affect the quality and distribution of food, with subsequent effects on food safety and nutrition.

Mental Health

Any changes in a person's physical health or surrounding environment can also have serious impacts on their mental health. In particular, experiencing an extreme weather event can cause stress and other mental health consequences, particularly when a person loses loved ones or their home.

- Individuals with mental illness are especially vulnerable to extreme heat; studies have found that having a pre-existing mental illness tripled the risk of death during heat waves. People taking medication for mental illness that makes it difficult to regulate their body temperature are particularly at risk.

- Even the perceived threat of climate change (for example from reading or watching news reports about climate change) can influence stress responses and mental health.

- Some groups of people are at higher risk for mental health impacts, such as children and older adults, pregnant and post-partum women, people with pre-existing mental illness, people with low incomes, and emergency workers.

Populations of Concern

Some groups of people are more vulnerable than others to health risks from climate change. Three factors contribute to vulnerability: *sensitivity*, which refers to the degree to which people or groups are affected by a stressor such as higher temperatures; *exposure*, which refers to physical contact between a person and a stressor; and *adaptive capacity*, which refers to an ability to adjust to or avoid potential hazards. For example, while older adults are sensitive to extreme heat, an older person living in an air-conditioned apartment won't be exposed as long as she stays indoors, and as long as she can afford to pay for the electricity to run the air conditioner. Her ability take these actions is a measure of her adaptive capacity.

Some populations are especially vulnerable to climate health risks due to particular sensitivities, high likelihood of exposure, low adaptive capacity, or combinations of these factors.

- Communities of color (including Indigenous communities as well as specific racial and ethnic groups), low income, immigrants, and limited English proficiency face disproportionate vulnerabilities due to a wide variety of factors, such as higher risk of exposure, socioeconomic and educational factors that affect their adaptive capacity, and a higher prevalence of medical conditions that affect their sensitivity.

- Children are vulnerable to many health risks due to biological sensitivities and more opportunities for exposure (due to activities such as playing outdoors). Pregnant women are vulnerable to heat waves and other extreme events, like flooding.

- Older adults are vulnerable to many of the impacts of climate change. They may have greater sensitivity to heat and contaminants, a higher prevalence of disability or preexisting medical conditions, or limited financial resources that make it difficult to adapt to impacts.

- Occupational groups, such as outdoor workers, paramedics, firefighters, and transportation workers, as well as workers in hot indoor work environments, will be especially vulnerable to extreme heat and exposure to vectorborne diseases.

- People with disabilities can be very vulnerable during extreme weather events, unless communities ensure that their emergency response plans specifically accommodate them.

- People with chronic medical conditions are typically vulnerable to extreme heat, especially if they are taking medications that make it difficult to regulate body temperature. Power outages can be particularly threatening for people reliant on certain medical equipment.

Other Health Impacts

Other linkages exist between climate change and human health. For example, changes in temperature and precipitation, as well as droughts and floods, will affect agricultural yields and production. In some regions of the world, these impacts may compromise food security and threaten human health through malnutrition, the spread of infectious diseases, and food poisoning. The worst of these effects are projected to occur in developing countries, among vulnerable populations. Declines in human health in other countries can affect the United States through trade, migration, and immigration and has implications for national security.

Although the impacts of climate change have the potential to affect human health in the United States and around the world, there is a lot we can do to prepare for and adapt to these changes—such as establishing early warning systems for heat waves and other extreme events, taking steps to reduce vulnerabilities among populations of concern, raising awareness among healthcare professionals, and ensuring that infrastructure is built to accommodate anticipated future changes in climate. Understanding the threats that climate change poses to human health is the first step in working together to lower risks and be prepared.

Climate Change and Agriculture

Climate change and agriculture are interrelated processes, both of which take place on a global scale. Climate change affects agriculture in a number of ways, including through changes in average temperatures, rainfall, and climate extremes (e.g., heat waves); changes in pests and diseases; changes in atmospheric carbon dioxide and ground-level ozone concentrations; changes in the nutritional quality of some foods; and changes in sea level.

Climate change is already affecting agriculture, with effects unevenly distributed across the world. Future climate change will likely negatively affect crop production in low latitude countries, while effects in northern latitudes may be positive or negative. Climate change will probably increase the risk of food insecurity for some vulnerable groups, such as the poor. Animal agriculture is also responsible for CO_2 greenhouse gas production and a percentage of the world's methane, and future land infertility, and the displacement of local species.

Agriculture contributes to climate change both by anthropogenic emissions of greenhouse gases and by the conversion of non-agricultural land such as forests into agricultural land. Agriculture, forestry and land-use change contributed around 20 to 25% of global annual emissions in 2010.

A range of policies can reduce the risk of negative climate change impacts on agriculture and greenhouse gas emissions from the agriculture sector.

Impact of Climate Change on Agriculture

Despite technological advances, such as improved varieties, genetically modified organisms, and irrigation systems, weather is still a key factor in agricultural productivity, as well as soil properties and natural communities. The effect of climate on agriculture is related to variabilities in local climates rather than in global climate patterns. The Earth's average surface temperature has increased by 1.5 °F (0.83 °C) since 1880. Consequently, in making an assessment agronomists must consider each local area.

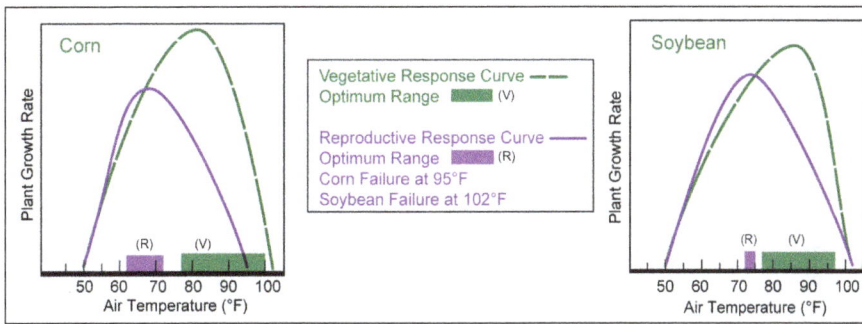

For each plant variety, there is an optimal temperature for vegetative growth, with growth dropping off as temperatures increase or decrease. Similarly, there is a range of temperatures at which a plant will produce seed. Outside of this range, the plant will not reproduce. As the graphs show, corn will fail to reproduce at temperatures above 95 °F (35 °C) and soybean above 102 °F (38.8 °C).

On the other hand, agricultural trade has grown in recent years, and now provides significant amounts of food, on a national level to major importing countries, as well as comfortable income to exporting ones. The international aspect of trade and security in terms of food implies the need to also consider the effects of climate change on a global scale.

A 2008 study published in *Science* suggested that, due to climate change, "southern Africa could lose more than 30% of its main crop, maize, by 2030. In South Asia losses of many regional staples, such as rice, millet and maize could top 10%".

The Intergovernmental Panel on Climate Change (IPCC) has produced several reports that have assessed the scientific literature on climate change. The IPCC Third Assessment Report, published in 2001, concluded that the poorest countries would be hardest hit, with reductions in crop yields in most tropical and sub-tropical regions due to decreased water availability, and new or changed insect pest incidence. In Africa and Latin America many rainfed crops are near their maximum temperature tolerance, so that yields are likely to fall sharply for even small climate changes; falls in agricultural productivity of up to 30% over the 21st century are projected. Marine life and the fishing industry will also be severely affected in some places.

Climate change induced by increasing greenhouse gases is likely to affect crops differently from region to region. For example, average crop yield is expected to drop down to 50% in Pakistan according to the Met Office scenario whereas corn production in Europe is expected to grow up to 25% in optimum hydrologic conditions.

More favourable effects on yield tend to depend to a large extent on realization of the potentially beneficial effects of carbon dioxide on crop growth and increase of efficiency in water use. Decrease in potential yields is likely to be caused by shortening of the growing period, decrease in water availability and poor vernalization.

In the long run, the climatic change could affect agriculture in several ways :

- Productivity, in terms of quantity and quality of crops.

- Agricultural practices, through changes of water use (irrigation) and agricultural inputs such as herbicides, insecticides and fertilizers.

- Environmental effects, in particular in relation of frequency and intensity of soil drainage (leading to nitrogen leaching), soil erosion, reduction of crop diversity.

- Rural space, through the loss and gain of cultivated lands, land speculation, land renunciation, and hydraulic amenities.

- Adaptation, organisms may become more or less competitive, as well as humans may develop urgency to develop more competitive organisms, such as flood resistant or salt resistant varieties of rice.

They are large uncertainties to uncover, particularly because there is lack of information on many specific local regions, and include the uncertainties on magnitude of climate change, the effects of technological changes on productivity, global food demands, and the numerous possibilities of adaptation.

Most agronomists believe that agricultural production will be mostly affected by the severity and pace of climate change, not so much by gradual trends in climate. If change is gradual, there may be enough time for biota adjustment. Rapid climate change, however, could harm agriculture in many countries, especially those that are already suffering from rather poor soil and climate conditions, because there is less time for optimum natural selection and adaption.

But much remains unknown about exactly how climate change may affect farming and food security, in part because the role of farmer behaviour is poorly captured by crop-climate models. For instance, Evan Fraser, a geographer at the University of Guelph in Ontario Canada, has conducted a number of studies that show that the socio-economic context of farming may play a huge role in determining whether a drought has a major, or an insignificant impact on crop production. In some cases, it seems that even minor droughts have big impacts on food security (such as what happened in Ethiopia in the early 1980s where a minor drought triggered a massive famine), versus cases where even relatively large weather-related problems were adapted to without much hardship. Evan Fraser combines socio-economic models along with climatic models to identify "vulnerability hotspots" One such study has identified US maize (corn) production as particularly vulnerable to climate change because it is expected to be exposed to worse droughts, but it does not have the socio-economic conditions that suggest farmers will adapt to these changing conditions. Other studies rely instead on projections of key agro-meteorological or agro-climate indices, such as growing season length, plant heat stress, or start of field operations, identified by land management stakeholders and that provide useful information on mechanisms driving climate change impact on agriculture.

Pest Insects and Climate Change

Global warming could lead to an increase in pest insect populations, harming yields of staple crops like wheat, soybeans, and corn. While warmer temperatures create longer growing seasons, and faster growth rates for plants, it also increases the metabolic rate and number of breeding cycles of insect populations. Insects that previously had only two breeding cycles per year could gain an additional cycle if warm growing seasons extend, causing a population boom. Temperate places and higher latitudes are more likely to experience a dramatic change in insect populations.

The University of Illinois conducted studies to measure the effect of warmer temperatures on soybean plant growth and Japanese beetle populations. Warmer temperatures and elevated CO_2 levels were simulated for one field of soybeans, while the other was left as a control. These studies found that the soybeans with elevated CO_2 levels grew much faster and had higher yields, but attracted Japanese beetles at a significantly higher rate than the control field. The beetles in the field with increased CO_2 also laid more eggs on the soybean plants and had longer lifespans, indicating the possibility of a rapidly expanding population. DeLucia projected that if the project were to continue, the field with elevated CO_2 levels would eventually show lower yields than that of the control field.

The increased CO_2 levels deactivated three genes within the soybean plant that normally create chemical defenses against pest insects. One of these defenses is a protein that blocks digestion of the soy leaves in insects. Since this gene was deactivated, the beetles were able to digest a much higher amount of plant matter than the beetles in the control field. This led to the observed longer lifespans and higher egg-laying rates in the experimental field.

There are a few proposed solutions to the issue of expanding pest populations. One proposed solution is to increase the number of pesticides used on future crops. This has the benefit of being relatively cost effective and simple, but may be ineffective. Many pest insects have been building up an immunity to these pesticides. Another proposed solution is to utilize biological control agents. This includes things like planting rows of native vegetation in between rows of crops. This solution is beneficial in its overall environmental impact. Not only are more native plants getting planted, but pest insects are no longer building up an immunity to pesticides. However, planting additional native plants requires more room, which destroys additional acres of public land. The cost is also much higher than simply using pesticides.

Plant Diseases and Climate Change

Although research is limited, research has shown that climate change may alter the developmental stages of pathogens that can affect crops. The biggest consequence of climate change on the dispersal of pathogens is that the geographical distribution of hosts and pathogens could shift, which would result in more crop losses. This could affect competition and recovery from disturbances of plants. It has been predicted that the effect of climate change will add a level of complexity to figuring out how to maintain sustainable agriculture.

Observed Impacts

Effects of regional climate change on agriculture have been limited. Changes in crop phenology provide important evidence of the response to recent regional climate change. Phenology is the study of natural phenomena that recur periodically, and how these phenomena relate to climate and seasonal changes. A significant advance in phenology has been observed for agriculture and forestry in large parts of the Northern Hemisphere.

Droughts have been occurring more frequently because of global warming and they are expected to become more frequent and intense in Africa, southern Europe, the Middle East, most of the Americas, Australia, and Southeast Asia. Their impacts are aggravated because of increased water demand, population growth, urban expansion, and environmental protection efforts in many areas. Droughts result in crop failures and the loss of pasture grazing land for livestock.

Banana farm.

As of the decade starting in 2010, many hot countries have thriving agricultural sectors.

Jalgaon district, India, has an average temperature which ranges from 20.2 °C in December to 29.8 °C in May, and an average precipitation of 750 mm/year. It produces bananas at a rate that would make it the world's seventh-largest banana producer if it were a country.

During the period 1990-2012, Nigeria had an average temperature which ranged from a low of 24.9 °C in January to a high of 30.4 °C in April. According to the Food and Agriculture Organization of the United Nations (FAO), Nigeria is by far the world's largest producer of yams, producing over 38 million tonnes in 2012. The second through 8th largest yam producers were all nearby African countries, with the largest non-African producer, Papua New Guinea, producing less than 1% of Nigerian production.

In 2013, according to the FAO, Brazil and India were by far the world's leading producers of sugarcane, with a combined production of over 1 billion tonnes, or over half of worldwide production.

In the summer of 2018, heat waves probably linked to climate change cause much lower than average yield in many parts of the world, especially in Europe. Depending on conditions during August, more crop failures could rise global food prices. losses are compared to those of 1945, the worst harvest in memory. last year was the third time in four years that global wheat, rice and maize production failed to meet demand, forcing governments and food companies to release stocks from storage. India last week released 50% of its food stocks. Lester Brown, the head of Worldwatch, an independent research organisation, predicted thatfood prices will rise in the next few months.

According to the UN report "Climate Change and Land: an IPCC special report on climate change, desertification, land degradation, sustainable land management, food security, and greenhouse gas fluxes in terrestrial ecosystems", food prices will rise by 80% by 2050 and food shortages are likely to occur. Some authors also suggested that the food shortages will probably affect poorer parts of the world far more than richer ones.

To prevent hunger, instability, new waves of climate refugees, international help will be needed to countries who will miss the money to buy enough food and for also for stopping conflicts.

At the beginning of the 21 century, floods probably linked to climate change shortened the planting season in the Midwest region in United States, what cause damage to the agriculture sector. In May 2019 the floods reduced the projected Corn yield from 15 billion bushels to 14.2.

Projections of Impacts

As part of the IPCC's Fourth Assessment Report, Schneider *et al.* projected the potential future effects of climate change on agriculture. With low to medium confidence, they concluded that for about a 1 to 3 °C global mean temperature increase (by 2100, relative to the 1990–2000 average level) there would be productivity decreases for some cereals in low latitudes, and productivity increases in high latitudes. In the IPCC Fourth Assessment Report, "low confidence" means that a particular finding has about a 2 out of 10 chance of being correct, based on expert judgement. "Medium confidence" has about a 5 out of 10 chance of being correct. Over the same time period, with medium confidence, global production potential was projected to:

- Increase up to around 3 °C,

- Very likely decrease above about 3 °C.

Most of the studies on global agriculture assessed by Schneider et al. had not incorporated a number of critical factors, including changes in extreme events, or the spread of pests and diseases. Studies had also not considered the development of specific practices or technologies to aid adaptation to climate change.

The US National Research Council assessed the literature on the effects of climate change on crop yields. US NRC stressed the uncertainties in their projections of changes in crop yields.

Writing in the journal Nature Climate Change, Matthew Smith and Samuel Myers estimated that food crops could see a reduction of protein, iron and zinc content in common food crops of 3 to 17%. This is the projected result of food grown under the expected atmospheric carbon-dioxide levels of 2050. Using data from the UN Food and Agriculture Organization as well as other public sources, the authors analyzed 225 different staple foods, such as wheat, rice, maize, vegetables, roots and fruits.

Their central estimates of changes in crop yields are shown above. Actual changes in yields may be above or below these central estimates. US NRC also provided an estimated the "likely" range of changes in yields. "Likely" means a greater than 67% chance of being correct, based on expert judgement. The likely ranges are summarized in the image descriptions of the two graphs.

Food Security

The IPCC Fourth Assessment Report also describes the impact of climate change on food security. Projections suggested that there could be large decreases in hunger globally by 2080, compared to the (then-current) 2006 level. Reductions in hunger were driven by projected social and economic development. For reference, the Food and Agriculture Organization has estimated that in 2006, the number of people undernourished globally was 820 million. Three scenarios without climate change (SRES A1, B1, B2) projected 100-130 million undernourished by the year 2080, while another scenario without climate change (SRES A2) projected 770 million undernourished. Based on an expert assessment of all of the evidence, these projections were thought to have about a 5-in-10 chance of being correct.

The same set of greenhouse gas and socio-economic scenarios were also used in projections that included the effects of climate change. Including climate change, three scenarios (SRES A1, B1, B2) projected 100-380 million undernourished by the year 2080, while another scenario with climate change (SRES A2) projected 740-1,300 million undernourished. These projections were thought to have between a 2-in-10 and 5-in-10 chance of being correct.

Projections also suggested regional changes in the global distribution of hunger. By 2080, sub-Saharan Africa may overtake Asia as the world's most food-insecure region. This is mainly due to projected social and economic changes, rather than climate change.

In South America, a phenomenon known as the El Nino Oscillation Cycle, between floods and drought on the Pacific Coast has made as much as a 35% difference in Global yields of wheat and grain.

Looking at the four key components of food security we can see the impact climate change has had. Food "Access to food is largely a matter of household and individual-level income and of capabilities and rights". Access has been affected by the thousands of crops being destroyed, how communities are dealing with climate shocks and adapting to climate change. Prices on food will rise due to the shortage of food production due to conditions not being favourable for crop production. Utilization is affected by floods and drought where water resources are contaminated, and the changing temperatures create vicious stages and phases of disease. Availability is affected by the contamination of the crops, as there will be no food process for the products of these crops as a result. Stability is affected through price ranges and future prices as some food sources are becoming scarce due to climate change, so prices will rise.

Individual Studies

Cline looked at how climate change might affect agricultural productivity in the 2080s. His study assumes that no efforts are made to reduce anthropogenic greenhouse gas emissions, leading to global warming of 3.3 °C above the pre-industrial level. He concluded that global agricultural productivity could be negatively affected by climate change, with the worst effects in developing countries.

Lobell et al. assessed how climate change might affect 12 food-insecure regions in 2030. The purpose of their analysis was to assess where adaptation measures to climate change should be prioritized. They found that without sufficient adaptation measures, South Asia and South Africa would likely suffer negative impacts on several crops which are important to large food insecure human populations.

Battisti and Naylor looked at how increased seasonal temperatures might affect agricultural productivity. Projections by the IPCC suggest that with climate change, high seasonal temperatures will become widespread, with the likelihood of extreme temperatures increasing through the second half of the 21st century. Battisti and Naylor concluded that such changes could have very serious effects on agriculture, particularly in the tropics. They suggest that major, near-term, investments in adaptation measures could reduce these risks.

"Climate change merely increases the urgency of reforming trade policies to ensure that global food security needs are met" said C. Bellmann, ICTSD Programmes Director. A study suggests that climate change could cause farm output in sub-Saharan Africa to decrease by 12% by 2080 - although in some African countries this figure could be as much as 60%, with agricultural exports declining by up to one fifth in others. Adapting to climate change could cost the agriculture sector $14bn globally a year, the study finds.

Regional Impacts

Africa

Agriculture is a particularly important sector in Africa, contributing towards livelihoods and economies across the continent. On Average, agriculture in Sub-Saharan Africa contributes 15% of the total GDP. Africa's geography makes it particularly vulnerable to climate change, and seventy per cent of the population rely on rain-fed agriculture for their livelihoods. Smallholder farms account for 80% of cultivated lands in sub-Saharan Africa. The Intergovernmental Panel on Climate Change (IPCC) projected that climate variability and change would severely compromise agricultural production and access to food. This projection was assigned "high confidence". Cropping systems, livestock and fisheries will be at greater risk of pest and diseases as a result of future climate change. Research program on Climate Change, Agriculture and Food Security (CCAFS) have identified that crop pests already account for approximately 1/6th of farm productivity losses. Similarly, climate change will accelerate the prevalence of pests and diseases and increase the occurrence of highly impactful events . The impacts of climate change on agricultural production in Africa will have serious implications for food security and livelihoods. Between 2014 and 2018, Africa had the highest levels of food insecurity in the world.

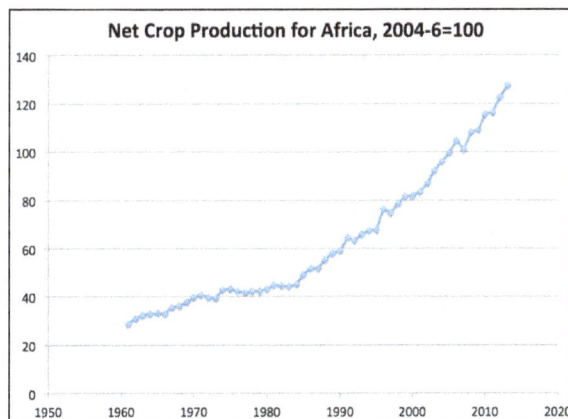

African crop production. Raw data from the United Nations.

In East Africa, climate change is anticipated to intensify the frequency and intensity of drought and flooding, which can have an adverse impact on the agricultural sector. Climate change will have varying effects on agricultural production in East Africa. Research from the International Food Policy Research Institute (IFPRI) suggest an increase in maize yields for most East Africa, but yield losses in parts of Ethiopia, Democratic Republic of Congo (DRC), Tanzania and northern Uganda. Projections of climate change are also anticipated to reduce the potential of the cultivated land to produce crops of high quantity and quality.

In Tanzania there is currently no clear signal in future climate projections for rainfall. However, there is a higher likelihood of intense future rainfall events. As of 2005, the net result was expected to be that 33% less maize—the country's staple crop—would be grown.

Climate change will exacerbate the vulnerability of the Agricultural sector in most Southern Afrcan countries which are already limited by poor infrastructure and a lag in technological inputs and innovation. Maize accounts for nearly half of the cultivated land in Southern Africa, and under future climate change, yields could decrease by 30% Temperatures increases also encourage a wide spread of weeds and pests.

Climate change will significantly affect agriculture in West Africa by increasing the variability in food production, access and availability. The region has already experienced a decrease in rainfall along the coasts of Nigeria, Sierra Leone, Guinea and Liberia. This has resulted in lower crop yield, causing farmers to seek new areas for cultivation. Staple crops such as maize, rice and sorghum will be impacted by low rainfall events with possible increase in food insecurity.

Higher rainfall intensity, prolonged dry spells and high temperatures are expected to negatively impact cassava, maize and bean production in Central Africa. Floods and erosion occurrence are expected to damage the already limited transportation infrastructure in the region leading to post harvest losses. Exportation of economic crops like coffee and cocoa are on the rise within the region but these crops are highly vulnerable to climate change . Conflicts and political instability have had an impact on agriculture contribution to the regional GDP and this impact will be exacerbated by climatic risks.

Asia

In East and Southeast Asia, IPCC projected that crop yields could increase up to 20% by the mid-21st century. In Central and South Asia, projections suggested that yields might decrease by up to 30%, over the same time period. These projections were assigned "medium confidence." Taken together, the risk of hunger was projected to remain very high in several developing countries.

More detailed analysis of rice yields by the International Rice Research Institute forecast 20% reduction in yields over the region per degree Celsius of temperature rise. Rice becomes sterile if exposed to temperatures above 35 degrees for more than one hour during flowering and consequently produces no grain.

A 2013 study by the International Crops Research Institute for the Semi-Arid Tropics (ICRISAT) aimed to find science-based, pro-poor approaches and techniques that would enable Asia's agricultural systems to cope with climate change, while benefitting poor and vulnerable farmers. The study's recommendations ranged from improving the use of climate information in local planning and strengthening weather-based agro-advisory services, to stimulating diversification of rural household incomes and providing incentives to farmers to adopt natural resource conservation measures to enhance forest cover, replenish groundwater and use renewable energy. A 2014 study found that warming had increased maize yields in the Heilongjiang region of China had increased by between 7 and 17% per decade as a result of rising temperatures.

Due to climate change, livestock production will be decreased in Bangladesh by diseases, scarcity of forage, heat stress and breeding strategies.

Australia

Hennessy *et al.* assessed the literature for Australia and New Zealand. They concluded that without further adaptation to climate change, projected impacts would likely be substantial: By 2030, production from agriculture and forestry was projected to decline over much of southern and eastern Australia, and over parts of eastern New Zealand; In New Zealand, initial benefits were projected close to major rivers and in western and southern areas. Hennessy *et al.* placed high confidence in these projections.

Europe

With high confidence, IPCC projected that in Southern Europe, climate change would reduce crop productivity. In Central and Eastern Europe, forest productivity was expected to decline. In Northern Europe, the initial effect of climate change was projected to increase crop yields.

Latin America

The major agricultural products of Latin American regions include livestock and grains, such as maize, wheat, soybeans, and rice. Increased temperatures and altered hydrological cycles are predicted to translate to shorter growing seasons, overall reduced biomass production, and lower grain yields. Brazil, Mexico and Argentina alone contribute 70-90% of the total agricultural production in Latin America. In these and other dry regions, maize production is expected to decrease. A study summarizing a number of impact studies of climate change on agriculture in Latin America indicated that wheat is expected to decrease in Brazil, Argentina and Uruguay. Livestock, which is the main agricultural product for parts of Argentina, Uruguay, southern Brazil, Venezuela, and Colombia is likely to be reduced. Variability in the degree of production decrease among different regions of Latin America is likely. For example, one 2003 study that estimated future maize production in Latin America predicted that by 2055 maize in eastern Brazil will have moderate changes while Venezuela is expected to have drastic decreases.

Suggested potential adaptation strategies to mitigate the impacts of global warming on agriculture in Latin America include using plant breeding technologies and installing irrigation infrastructure.

Climate Justice and Subsistence Farmers

Several studies that investigated the impacts of climate change on agriculture in Latin America suggest that in the poorer countries of Latin America, agriculture composes the most important economic sector and the primary form of sustenance for small farmers. Maize is the only grain still produced as a sustenance crop on small farms in Latin American nations. Scholars argue that the projected decrease of this grain and other crops will threaten the welfare and the economic development of subsistence communities in Latin America. Food security is of particular concern to rural areas that have weak or non-existent food markets to rely on in the case food shortages.

According to scholars who considered the environmental justice implications of climate change, the expected impacts of climate change on subsistence farmers in Latin America and other developing regions are unjust for two reasons. First, subsistence farmers in developing countries, including those in Latin America are disproportionately vulnerable to climate change Second, these nations were the least responsible for causing the problem of anthropogenic induced climate.

According to researchers John F. Morton and T. Roberts, disproportionate vulnerability to climate disasters is socially determined. For example, socioeconomic and policy trends affecting small-holder and subsistence farmers limit their capacity to adapt to change. According to W. Baethgen who studied the vulnerability of Latin American agriculture to climate change, a history of policies and economic dynamics has negatively impacted rural farmers. During the 1950s and through the 1980s, high inflation and appreciated real exchange rates reduced the value of agricultural exports. As a result, farmers in Latin America received lower prices for their products compared to world market prices. Following these outcomes, Latin American policies and national crop programs aimed to stimulate agricultural intensification. These national crop programs benefitted larger commercial farmers more. In the 1980s and 1990s low world market prices for cereals and live-stock resulted in decreased agricultural growth and increased rural poverty.

In the book, Fairness in Adaptation to Climate Change, the authors describe the global injustice of climate change between the rich nations of the north, who are the most responsible for global warming and the southern poor countries and minority populations within those countries who are most vulnerable to climate change impacts.

Adaptive planning is challenged by the difficulty of predicting local scale climate change impacts. An expert that considered opportunities for climate change adaptation for rural communities argues that a crucial component to adaptation should include government efforts to lessen the effects of food shortages and famines. This researcher also claims that planning for equitable adaptation and agricultural sustainability will require the engagement of farmers in decision making processes.

North America

A number of studies have been produced which assess the impacts of climate change on agriculture in North America. The IPCC Fourth Assessment Report of agricultural impacts in the region cites 26 different studies. With high confidence, IPCC projected that over the first few decades of this century, moderate climate change would increase aggregate yields of rain-fed agriculture by 5–20%, but with important variability among regions. Major challenges were projected for crops that are near the warm end of their suitable range or which depend on highly utilized water resources.

Droughts are becoming more frequent and intense in arid and semiarid western North America as temperatures have been rising, advancing the timing and magnitude of spring snow melt floods and reducing river flow volume in summer. Direct effects of climate change include increased heat and water stress, altered crop phenology, and disrupted symbiotic interactions. These effects may be exacerbated by climate changes in river flow, and the combined effects are likely to reduce the abundance of native trees in favor of non-native herbaceous and drought-tolerant competitors, reduce the habitat quality for many native animals, and slow litter decomposition and nutrient cycling. Climate change effects on human water demand and irrigation may intensify these effects.

The US Global Change Research Program assessed the literature on the impacts of climate change on agriculture in the United States, finding that many crops will benefit from increased atmospheric CO_2 concentrations and low levels of warming, but that higher levels of warming will negatively affect growth and yields; that extreme weather events will likely reduce crop yields; that weeds, diseases and insect pests will benefit from warming, and will require additional pest and weed control; and that increasing CO_2 concentrations will reduce the land's ability to supply adequate livestock feed, while increased heat, disease, and weather extremes will likely reduce livestock productivity.

Polar Regions

Anisimov *et al.* assessed the literature for the polar region (Arctic and Antarctica). With medium confidence, they concluded that the benefits of a less severe climate were dependent on local conditions. One of these benefits was judged to be increased agricultural and forestry opportunities.

The Guardian reported on how climate change had affected agriculture in Iceland. Rising temperatures had made the widespread sowing of barley possible, which had been untenable twenty years ago. Some of the warming was due to a local (possibly temporary) effect via ocean currents from the Caribbean, which had also affected fish stocks.

Small Islands

In a literature assessment, Mimura *et al.* concluded that on small islands, subsistence and commercial agriculture would very likely be adversely affected by climate change. This projection was assigned "high confidence."

Climate Change Impacts Affecting Poverty Alleviation

Researchers at the Overseas Development Institute (ODI) have investigated the potential impacts climate change could have on agriculture, and how this would affect attempts at alleviating poverty in the developing world. They argued that the effects from moderate climate change are likely to be mixed for developing countries. However, the vulnerability of the poor in developing countries to short term impacts from climate change, notably the increased frequency and severity of adverse weather events is likely to have a negative impact. This, they say, should be taken into account when defining agricultural policy.

Crop Development Models

Models for climate behavior are frequently inconclusive. In order to further study effects of global warming on agriculture, other types of models, such as crop development models, yield prediction, quantities of water or fertilizer consumed, can be used. Such models condense the knowledge accumulated of the climate, soil, and effects observed of the results of various agricultural practices. They thus could make it possible to test strategies of adaptation to modifications of the environment.

Because these models are necessarily simplifying natural conditions (often based on the assumption that weeds, disease and insect pests are controlled), it is not clear whether the results they give will have an *in-field* reality. However, some results are partly validated with an increasing number of experimental results.

Other models, such as *insect and disease development* models based on climate projections are also used (for example simulation of aphid reproduction or septoria (cereal fungal disease) development).

Scenarios are used in order to estimate climate changes effects on crop development and yield. Each scenario is defined as a set of meteorological variables, based on generally accepted projections. For example, many models are running simulations based on doubled carbon dioxide projections, temperatures raise ranging from 1 °C up to 5 °C, and with rainfall levels an increase or decrease of 20%. Other parameters may include humidity, wind, and solar activity. Scenarios of crop models are testing farm-level adaptation, such as sowing date shift, climate adapted species (vernalisation need, heat and cold resistance), irrigation and fertilizer adaptation, resistance to disease. Most developed models are about wheat, maize, rice and soybean.

Temperature Potential Effect on Growing Period

Duration of crop growth cycles are above all, related to temperature. An increase in temperature will speed up development. In the case of an annual crop, the duration between sowing and harvesting will shorten (for example, the duration in order to harvest corn could shorten between one and four weeks). The shortening of such a cycle could have an adverse effect on productivity because senescence would occur sooner.

Effect of Elevated Carbon Dioxide on Crops

Elevated atmospheric carbon dioxide affects plants in a variety of ways. Elevated CO_2 increases crop yields and growth through an increase in photosynthetic rate, and it also decreases water loss as a result of stomatal closing The growth response is greatest in C_3 plants, C_4 plants, are also enhanced but to a lesser extent, and CAM Plants are the least enhanced species. The stoma in these "CAM plant" stores remain shut all day to reduce exposure. rapidly rising levels of carbon dioxide in the atmosphere affect plants' absorption of nitrogen, which is the nutrient that restricts crop growth in most terrestrial ecosystems. Today's concentration of 400 ppm plants are relatively starved for nutrition. The optimum level of CO_2 for plant growth is about 5 times higher. Increased mass of CO_2 increases photosynthesis, this CO_2 potentially stunts the growth of the plant. It limit's the reduction that crops lose through transpiration.

Effect of Drought Stress on Crops

Increase in global temperatures will cause an increase in evaporation rates and annual evaporation levels. Increased evaporation will lead to an increase in storms in some areas, while leading to accelerated drying of other areas. These storm impacted areas will likely experience increased levels of precipitation and increased flood risks, while areas outside of the storm track will experience less precipitation and increased risk of droughts. Water stress effects plant development and quality in a variety of ways first off drought can cause poor germination and impaired seedling development in plants. At the same time plant growth relies on cellular division, cell enlargement, and differentiation. Drought stress impairs mitosis and cell elongation via loss of turgor pressure which results in poor growth. Development of leaves is also dependent upon turgor pressure, concentration of nutrients, and carbon assimilates all of which are reduced by drought conditions, thus drought stress lead to a decrease in leaf size and number. Plant height, biomass, leaf size and stem girth has been shown to decrease in Maize under water limiting conditions. Crop yield is also negatively effected by drought stress, the reduction in crop yield results from a decrease in photosynthetic rate, changes in leaf development, and altered allocation of resources all due to drought stress. Crop plants exposed to drought stress suffer from reductions in leaf water potential and transpiration rate, however water-use efficiency has been shown to increase in some crop plants such as wheat while decreasing in others such as potatoes. Plants need water for the uptake of nutrients from the soil, and for the transport of nutrients throughout the plant, drought conditions limit these functions leading to stunted growth. Drought stress also causes a decrease in photosynthetic activity in plants due to the reduction of photosynthetic tissues, stomatal closure, and reduced performance of photosynthetic machinery. This reduction in photosynthetic activity contributes to the reduction in plant growth and yields. Another factor influencing reduced plant growth and yields include the allocation of resources; following drought stress plants will allocate more resources to roots to aid in water uptake increasing root growth and reducing the growth of other plant parts while decreasing yields.

Effect on Quality

According to the IPCC's TAR, "The importance of climate change impacts on grain and forage quality emerges from new research. For rice, the amylose content of the grain—a major determinant of cooking quality—is increased under elevated CO_2". Cooked rice grain from plants grown in high-CO_2 environments would be firmer than that from today's plants. However, concentrations of iron and zinc, which are important for human nutrition, would be lower. Moreover, the protein content of the grain decreases under combined increases of temperature and CO_2. Studies using FACE have shown that increases in CO_2 lead to decreased concentrations of micronutrients in crop plants, including decreased B vitamins in rice. This may have knock-on effects on other parts of ecosystems as herbivores will need to eat more food to gain the same amount of protein.

Studies have shown that higher CO_2 levels lead to reduced plant uptake of nitrogen (and a smaller number showing the same for trace elements such as zinc) resulting in crops with lower nutritional value. This would primarily impact on populations in poorer countries less able to compensate by eating more food, more varied diets, or possibly taking supplements.

Reduced nitrogen content in grazing plants has also been shown to reduce animal productivity in sheep, which depend on microbes in their gut to digest plants, which in turn depend on nitrogen intake. Because of the lack of water available to crops in warmer countries they struggle to survive as they suffer from dehydration, taking into account the increasing demand for water outside of agriculture as well as other agricultural demands.

Effect of Hail

In North America, fewer hail days will occur overall due to climate change, but storms with larger hail might become more common (including hail that is larger than 1.6 inch). Hail that is larger than 1.6 inch can quite easily break (glass) greenhouses.

Agricultural Surfaces and Climate Changes

Climate change may increase the amount of arable land in high-latitude region by reduction of the amount of frozen lands. A 2005 study reports that temperature in Siberia has increased three degree Celsius in average since 1960 (much more than the rest of the world). However, reports about the impact of global warming on Russian agriculture indicate conflicting probable effects : while they expect a northward extension of farmable lands, they also warn of possible productivity losses and increased risk of drought.

Sea levels are expected to get up to one meter higher by 2100, though this projection is disputed. A rise in the sea level would result in an agricultural land loss, in particular in areas such as South East Asia. Erosion, submergence of shorelines, salinity of the water table due to the increased sea levels, could mainly affect agriculture through inundation of low-lying lands.

Low-lying areas such as Bangladesh, India and Vietnam will experience major loss of rice crop if sea levels rise as expected by the end of the century. Vietnam for example relies heavily on its southern tip, where the Mekong Delta lies, for rice planting. Any rise in sea level of no more than a meter will drown several km² of rice paddies, rendering Vietnam incapable of producing its main staple and export of rice.

Erosion and Fertility

The warmer atmospheric temperatures observed over the past decades are expected to lead to a more vigorous hydrological cycle, including more extreme rainfall events. Erosion and soil degradation is more likely to occur. Soil fertility would also be affected by global warming. Increased erosion in agricultural landscapes from anthropogenic factors can occur with losses of up to 22% of soil carbon in 50 years. However, because the ratio of soil organic carbon to nitrogen is mediated by soil biology such that it maintains a narrow range, a doubling of soil organic carbon is likely to imply a doubling in the storage of nitrogen in soils as organic nitrogen, thus providing higher available nutrient levels for plants, supporting higher yield potential. The demand for imported fertilizer nitrogen could decrease, and provide the opportunity for changing costly fertilisation strategies.

Due to the extremes of climate that would result, the increase in precipitations would probably result in greater risks of erosion, whilst at the same time providing soil with better hydration, according to the intensity of the rain. The possible evolution of the organic matter in the soil is a highly contested issue: while the increase in the temperature would induce a greater rate in the production of minerals, lessening the soil organic matter content, the atmospheric CO_2 concentration would tend to increase it.

Potential Effects of Global Climate Change on Pests, Diseases and Weeds

A very important point to consider is that weeds would undergo the same acceleration of cycle as cultivated crops, and would also benefit from carbonaceous fertilization. Since most weeds are C3 plants, they are likely to compete even more than now against C4 crops such as corn. However, on the other hand, some results make it possible to think that weedkillers could increase in effectiveness with the temperature increase.

Global warming would cause an increase in rainfall in some areas, which would lead to an increase of atmospheric humidity and the duration of the wet seasons. Combined with higher temperatures, these could favor the development of fungal diseases. Similarly, because of higher temperatures and humidity, there could be an increased pressure from insects and disease vectors.

Glacier Retreat and Disappearance

The continued retreat of glaciers will have a number of different quantitative impacts. In the areas that are heavily dependent on water runoff from glaciers that melt during the warmer summer months, a continuation of the current retreat will eventually deplete the glacial ice and substantially reduce or eliminate runoff. A reduction in runoff will affect the ability to irrigate crops and will reduce summer stream flows necessary to keep dams and reservoirs replenished.

Approximately 2.4 billion people live in the drainage basin of the Himalayan rivers. India, China, Pakistan, Afghanistan, Bangladesh, Nepal and Myanmar could experience floods followed by severe droughts in coming decades. In India alone, the Ganges provides water for drinking and farming for more than 500 million people. The west coast of North America, which gets much of its water from glaciers in mountain ranges such as the Rocky Mountains and Sierra Nevada, also would be affected.

Ozone and UV-B

Some scientists think agriculture could be affected by any decrease in stratospheric ozone, which could increase biologically dangerous ultraviolet radiation B. Excess ultraviolet radiation B can directly affect plant physiology and cause massive amounts of mutations, and indirectly through changed pollinator behavior, though such changes are not simple to quantify. However, it has not yet been ascertained whether an increase in greenhouse gases would decrease stratospheric ozone levels.

In addition, a possible effect of rising temperatures is significantly higher levels of ground-level ozone, which would substantially lower yields.

ENSO Effects on Agriculture

ENSO (El Niño Southern Oscillation) will affect monsoon patterns more intensely in the future as climate change warms up the ocean's water. Crops that lie on the equatorial belt or under the tropical Walker circulation, such as rice, will be affected by varying monsoon patterns and more unpredictable weather. Scheduled planting and harvesting based on weather patterns will become less effective.

Areas such as Indonesia where the main crop consists of rice will be more vulnerable to the increased intensity of ENSO effects in the future of climate change. University of Washington professor, David Battisti, researched the effects of future ENSO patterns on the Indonesian rice agriculture using [IPCC]'s 2007 annual report and 20 different logistical models mapping out climate factors such as wind pressure, sea-level, and humidity, and found that rice harvest will experience a decrease in yield. Bali and Java, which holds 55% of the rice yields in Indonesia, will be likely to experience 9–10% probably of delayed monsoon patterns, which prolongs the hungry season. Normal planting of rice crops begin in October and harevest by January. However, as climate change affects ENSO and consequently delays planting, harvesting will be late and in drier conditions, resulting in less potential yields.

Mitigation and Adaptation

In Developed Countries

Several mitigation measures for use in developed countries have been proposed:

- Breeding more resilient crop varieties, and diversification of crop species,
- Using Improved agroforestry species,
- Capture and retention of rainfall, and use of improved irrigation practices,
- Increasing forest cover and Agroforestry,
- Use Of emerging water harvesting techniques (such as contour trenching, etc.).

In Developing Countries

The Intergovernmental Panel on Climate Change (IPCC) has reported that agriculture is responsible for over a quarter of total global greenhouse gas emissions. Given that agriculture's share in global gross domestic product (GDP) is about 4%, these figures suggest that agriculture is highly greenhouse gas intensive. Innovative agricultural practices and technologies can play a role in climate change mitigation and adaptation. This adaptation and mitigation potential is nowhere more pronounced than in developing countries where agricultural productivity remains low; poverty, vulnerability and food insecurity remain high; and the direct effects of climate change are expected to be especially harsh. Creating the necessary agricultural technologics and harnessing them to enable developing countries to adapt their agricultural systems to changing climate will require innovations in policy and institutions as well. In this context, institutions and policies are important at multiple scales.

Travis Lybbert and Daniel Sumner suggest six policy principles:

- The best policy and institutional responses will enhance information flows, incentives and flexibility.
- Policies and institutions that promote economic development and reduce poverty will often improve agricultural adaptation and may also pave the way for more effective climate change mitigation through agriculture.
- Business as usual among the world's poor is not adequate.
- Existing technology options must be made more available and accessible without overlooking complementary capacity and investments.
- Adaptation and mitigation in agriculture will require local responses, but effective policy responses must also reflect global impacts and inter-linkages.
- Trade will play a critical role in both mitigation and adaptation, but will itself be shaped importantly by climate change.

The Agricultural Model Intercomparison and Improvement Project (AgMIP) was developed in 2010 to evaluate agricultural models and intercompare their ability to predict climate impacts. In sub-Saharan Africa and South Asia, South America and East Asia, AgMIP regional research teams (RRTs) are conducting integrated assessments to improve understanding of agricultural impacts of climate change (including biophysical and economic impacts) at national and regional scales. Other AgMIP initiatives include global gridded modeling, data and information technology (IT) tool development, simulation of crop pests and diseases, site-based crop-climate sensitivity studies, and aggregation and scaling.

Impact of Agriculture on Climate Change

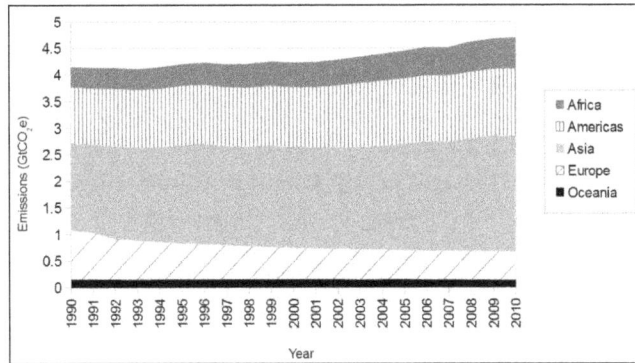

Greenhouse gas emissions from agriculture.

The agricultural sector is a driving force in the gas emissions and land use effects thought to cause climate change. In addition to being a significant user of land and consumer of fossil fuel, agriculture contributes directly to greenhouse gas emissions through practices such as rice production and the raising of livestock; according to the Intergovernmental Panel on Climate Change, the three main causes of the increase in greenhouse gases observed over the past 250 years have been fossil fuels, land use, and agriculture.

Land Use

Agriculture contributes to greenhouse gas increases through land use in four main ways:

- CO_2 releases linked to deforestation,
- Methane releases from rice cultivation,
- Methane releases from enteric fermentation in cattle,
- Nitrous oxide releases from fertilizer application.

Together, these agricultural processes comprise 54% of methane emissions, roughly 80% of nitrous oxide emissions, and virtually all carbon dioxide emissions tied to land use.

The planet's major changes to land cover since 1750 have resulted from deforestation in temperate regions: when forests and woodlands are cleared to make room for fields and pastures, the albedo of the affected area increases, which can result in either warming or cooling effects, depending on local conditions. Deforestation also affects regional carbon reuptake, which can result in increased concentrations of CO_2, the dominant greenhouse gas. Land-clearing methods such as slash and burn compound these effects by burning biomatter, which directly releases greenhouse gases and particulate matter such as soot into the air.

Livestock

Livestock and livestock-related activities such as deforestation and increasingly fuel-intensive farming practices are responsible for over 18% of human-made greenhouse gas emissions, including:

- 9% of global carbon dioxide emissions,

- 35–40% of global methane emissions (chiefly due to enteric fermentation and manure),

- 64% of global nitrous oxide emissions (chiefly due to fertilizer use).

Livestock activities also contribute disproportionately to land-use effects, since crops such as corn and alfalfa are cultivated in order to feed the animals.

In 2010, enteric fermentation accounted for 43% of the total greenhouse gas emissions from all agricultural activity in the world. The meat from ruminants has a higher carbon equivalent footprint than other meats or vegetarian sources of protein based on a global meta-analysis of lifecycle assessment studies. Methane production by animals, principally ruminants, is estimated 15-20% global production of methane.

Worldwide, livestock production occupies 70% of all land used for agriculture, or 30% of the land surface of the Earth. The way livestock is grazed also decides the fertility of the land in the future, not circulating grazing can lead to unhealthy soil and the expansion of livestock farms effects the habitats of local animals and has led to the drop in population of many local species from being displaced.

Climate Change Mitigation

Climate change mitigation consists of actions to limit the magnitude or rate of long-term global warming and its related effects. Climate change mitigation generally involves reductions in human (anthropogenic) emissions of greenhouse gases (GHGs). Mitigation may also be achieved by increasing the capacity of carbon sinks, e.g., through reforestation. Mitigation policies can substantially reduce the risks associated with human-induced global warming.

According to the IPCC's 2014 assessment report, "Mitigation is a public good; climate change is a case of the 'tragedy of the commons'. Effective climate change mitigation will not be achieved if each agent (individual, institution or country) acts independently in its own selfish interest suggesting the need for collective action. Some adaptation actions, on the other hand, have characteristics of a private good as benefits of actions may accrue more directly to the individuals, regions, or countries that undertake them, at least in the short term. Nevertheless, financing such adaptive activities remains an issue, particularly for poor individuals and countries."

Examples of mitigation include reducing energy demand by increasing energy efficiency, phasing out fossil fuels by switching to low-carbon energy sources, and removing carbon dioxide from Earth's atmosphere. for example, through improved building insulation. Another approach to climate change mitigation is climate engineering.

Most countries are parties to the United Nations Framework Convention on Climate Change

(UNFCCC). The ultimate objective of the UNFCCC is to stabilize atmospheric concentrations of GHGs at a level that would prevent dangerous human interference of the climate system. Scientific analysis can provide information on the impacts of climate change, but deciding which impacts are dangerous requires value judgments.

In 2010, Parties to the UNFCCC agreed that future global warming should be limited to below 2.0 °C (3.6 °F) relative to the pre-industrial level. With the Paris Agreement of 2015 this was confirmed, but was revised with a new target laying down "parties will do the best" to achieve warming below 1.5 °C. The current trajectory of global greenhouse gas emissions does not appear to be consistent with limiting global warming to below 1.5 or 2 °C. Other mitigation policies have been proposed, some of which are more stringent or modest than the 2 °C limit. In 2019, after 2 years of research, scientists from Australia, and Germany presented the "One Earth Climate Model" showing how temperature increase can be limited to 1.5 °C for 1.7 trillion dollars a year.

Greenhouse Gas Concentrations and Stabilization

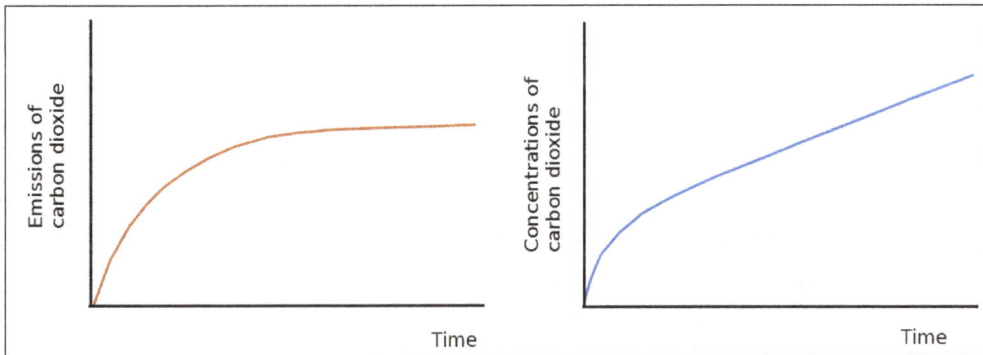

Stabilizing CO_2 emissions at their present level would not stabilize its concentration in the atmosphere.

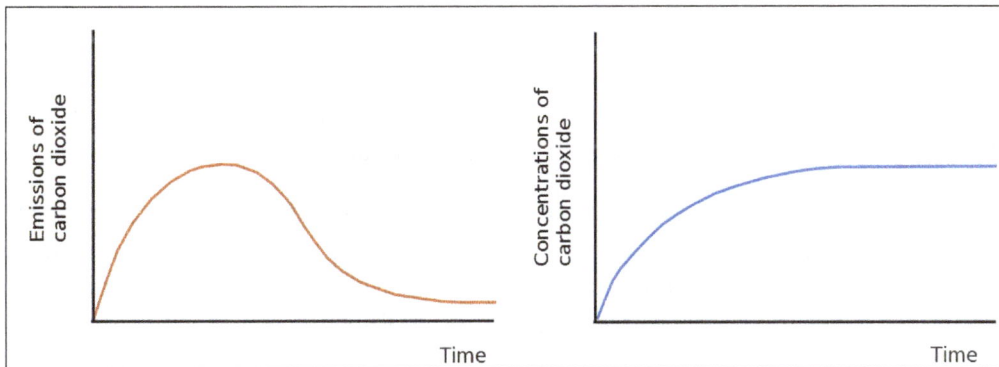

Stabilizing the atmospheric concentration of CO_2 at a constant level would require emissions to be effectively eliminated.

One of the issues often discussed in relation to climate change mitigation is the stabilization of greenhouse gas concentrations in the atmosphere. The United Nations Framework Convention on Climate Change (UNFCCC) has the ultimate objective of preventing "dangerous" anthropogenic (i.e., human) interference of the climate system. As is stated in Article 2 of the Convention, this requires that greenhouse gas (GHG) concentrations are stabilized in the atmosphere at a level where ecosystems can adapt naturally to climate change, food production is not threatened, and economic development can proceed in a sustainable fashion.

There are a number of anthropogenic greenhouse gases. These include carbon dioxide (chemical formula: CO_2), methane (CH_4), nitrous oxide (N_2O), and a group of gases referred to as halocarbons. Another greenhouse gas, water vapor, has also risen as an indirect result of human activities. The emissions reductions necessary to stabilize the atmospheric concentrations of these gases varies. CO_2 is the most important of the anthropogenic greenhouse gases.

There is a difference between stabilizing CO_2 emissions and stabilizing atmospheric concentrations of CO_2. Stabilizing emissions of CO_2 at current levels would not lead to a stabilization in the atmospheric concentration of CO_2. In fact, stabilizing emissions at current levels would result in the atmospheric concentration of CO_2 continuing to rise over the 21st century and beyond.

The reason for this is that human activities are adding CO_2 to the atmosphere faster than natural processes can remove it. This is analogous to a flow of water into a bathtub. So long as the tap runs water (analogous to the emission of carbon dioxide) into the tub faster than water escapes through the plughole (analogous to the natural removal of carbon dioxide from the atmosphere) the level of water in the tub (analogous to the concentration of carbon dioxide in the atmosphere) will continue to rise.

Stabilizing atmospheric CO_2 concentrations would require anthropogenic CO_2 emissions to be reduced by 80% relative to the peak emissions level. An 80% reduction in emissions would stabilize CO_2 concentrations for around a century, but even greater reductions would be required beyond this. Other research has found that, after leaving room for emissions for food production for 9 billion people and to keep the global temperature rise below 2 °C, emissions from energy production and transport will have to peak almost immediately in the developed world and decline at ca. 10% per annum until zero emissions are reached around 2030. In developing countries energy and transport emissions would have to peak by 2025 and then decline similarly.

Stabilizing the atmospheric concentration of the other greenhouse gasses humans emit also depends on how fast their emissions are added to the atmosphere, and how fast the GHGs are removed. Stabilization for these gases is described in the later section on non-CO_2 GHGs.

In 2018 an international team of scientist published research saying that the current mitigation policy in Paris Agreement is insufficient to limit the temperature rise to 2 degrees. They say that even if all the current pledges will be accomplished there is a chance for a 4.5 degree temperature rise in decades. To preventing that, restoration of natural Carbon sinks, Carbon dioxide removal, changes in society and values will be necessary.

Projections

Projections of future greenhouse gas emissions are highly uncertain. In the absence of policies to mitigate climate change, GHG emissions could rise significantly over the 21st century.

Numerous assessments have considered how atmospheric GHG concentrations could be stabilized. The lower the desired stabilization level, the sooner global GHG emissions must peak and decline. GHG concentrations are unlikely to stabilize this century without major policy changes.

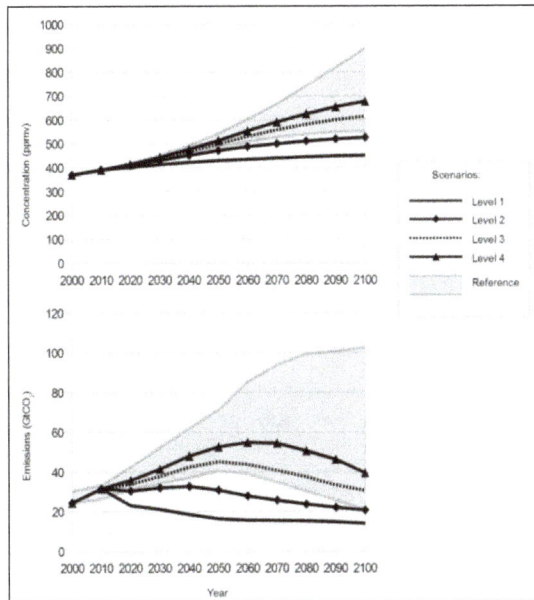

Projected carbon dioxide emissions and atmospheric concentrations over
the 21st century for reference and mitigation scenarios.

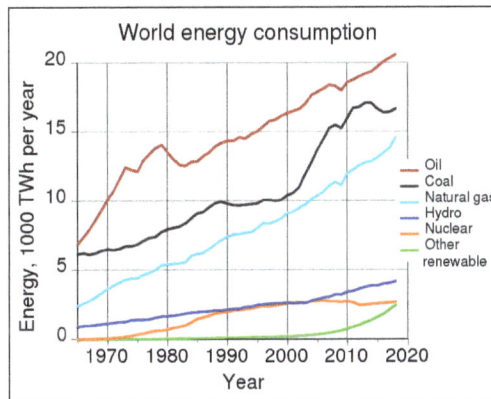

Rate of world energy usage per day.

Every fossil fuel source has increased in large amounts between 1970 and 2010, dominating all other energy sources. Hydroelectricity has increased at a slow steady rate over this same period, nuclear entered a period of rapid growth between 1970 and 1990 before leveling off. Other renewables, between 2000 and 2010 have, having started from a low usage rate, began to enter into a period of rapid growth. 1000 TWh=1 PWh.

Energy Consumption by Power Source

To create lasting climate change mitigation, the replacement of high carbon emission intensity power sources, such as conventional fossil fuels—oil, coal, and natural gas—with low-carbon power sources is required. Fossil fuels supply humanity with the vast majority of our energy demands, and at a growing rate. In 2012 the IEA noted that coal accounted for half the increased energy use of the prior decade, growing faster than all renewable energy sources. Both hydroelectricity and nuclear power together provide the majority of the generated low-carbon power fraction of global total power consumption.

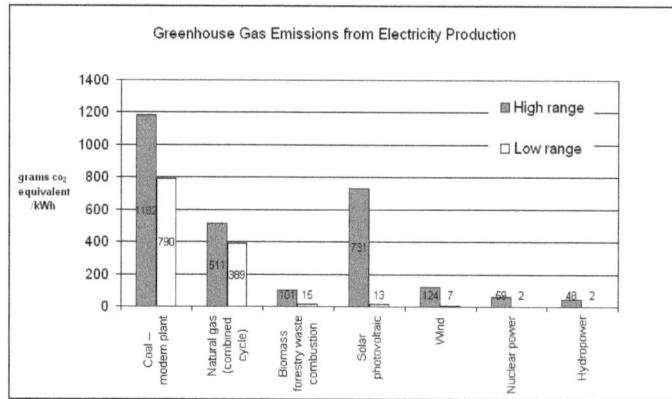

Greenhouse Gas Emissions from Electricity Production

"Hydropower-Internalised Costs and Externalised Benefits"; Frans H. Koch; International Energy Agency (IEA)-Implementing Agreement for Hydropower Technologies and Programmes; 2000.

Fuel type	Average total global power consumption in TW		
	1980	2004	2006
Oil	4.38	5.58	5.74
Gas	1.80	3.45	3.61
Coal	2.34	3.87	4.27
Hydroelectric	0.60	0.93	1.00
Nuclear power	0.25	0.91	0.93
Geothermal, wind, solar energy, wood	0.02	0.13	0.16
Total	9.48	15.0	15.8

Change and use of energy, by source, in units of (PWh) in that year				
	Fossil	Nuclear	All renewables	Total
1990	83.374	6.113	13.082	102.569
2000	94.493	7.857	15.337	117.687
2008	117.076	8.283	18.492	143.851
Change 2000–2008	22.583	0.426	3.155	26.164

Methods and Means

This graph shows the projected contribution of various energy sources to world primary electricity consumption (PEC). It is based on a climate change mitigation scenario, in which GHG emissions are substantially reduced over the 21st century. In the scenario, emission reductions are achieved using a portfolio of energy sources, as well as reductions in energy demand. Also available in grey-scale.

Assessments often suggest that GHG emissions can be reduced using a portfolio of low-carbon technologies. At the core of most proposals is the reduction of greenhouse gas (GHG) emissions through reducing energy waste and switching to low-carbon power sources of energy. As the cost of reducing GHG emissions in the electricity sector appears to be lower than in other sectors, such as in the transportation sector, the electricity sector may deliver the largest proportional carbon reductions under an economically efficient climate policy.

Economic tools can be useful in designing climate change mitigation policies." "While the limitations of economics and social welfare analysis, including cost–benefit analysis, are widely documented, economics nevertheless provides useful tools for assessing the pros and cons of taking, or not taking, action on climate change mitigation, as well as of adaptation measures, in achieving competing societal goals. Understanding these pros and cons can help in making policy decisions on climate change mitigation and can influence the actions taken by countries, institutions and individuals.

Other frequently discussed means include efficiency, public transport, increasing fuel economy in automobiles (which includes the use of electric hybrids), charging plug-in hybrids and electric cars by low-carbon electricity, making individual changes, and changing business practices. Many fossil fuel driven vehicles can be converted to use electricity, the US has the potential to supply electricity for 73% of light duty vehicles (LDV), using overnight charging. The US average CO_2 emissions for a battery-electric car is 180 grams per mile vs 430 grams per mile for a gasoline car. The emissions would be displaced away from street level, where they have "high human-health implications. Increased use of electricity "generation for meeting the future transportation load is primarily fossil-fuel based", mostly natural gas, followed by coal, but could also be met through nuclear, tidal, hydroelectric and other sources.

A range of energy technologies may contribute to climate change mitigation. These include nuclear power and renewable energy sources such as biomass, hydroelectricity, wind power, solar power, geothermal power, ocean energy, and; the use of carbon sinks, and carbon capture and storage. For example, Pacala and Socolow of Princeton have proposed a 15 part program to reduce CO_2 emissions by 1 billion metric tons per year – or 25 billion tons over the 50-year period using today's technologies as a type of global warming game.

Another consideration is how future socioeconomic development proceeds. Development choices (or "pathways") can lead differences in GHG emissions. Political and social attitudes may affect how easy or difficult it is to implement effective policies to reduce emissions.

Demand Side Management

Lifestyle and Behavior

The IPCC Fifth Assessment Report emphasises that behaviour, lifestyle, and cultural change have a high mitigation potential in some sectors, particularly when complementing technological and structural change. In general, higher consumption lifestyles have a greater environmental impact. Several scientific studies have shown that when people, especially those living in developed countries but more generally including all countries, wish to reduce their carbon footprint, there are four key "high-impact" actions they can take:

1. Not having an additional child (58.6 tonnes CO_2-equivalent emission reductions per year),

2. Living car-free (2.4 tonnes CO_2),

3. Avoiding one round-trip transatlantic flight (1.6 tonnes),

4. Eating a plant-based diet (0.8 tonnes).

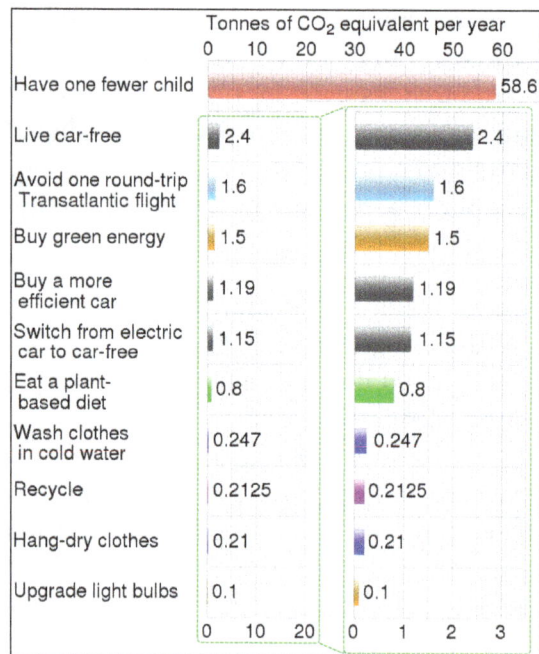

Reduction in one's carbon footprint for various actions.

These appear to differ significantly from the popular advice for "greening" one's lifestyle, which seem to fall mostly into the "low-impact" category: Replacing a typical car with a hybrid (0.52 tonnes); Washing clothes in cold water (0.25 tonnes); Recycling (0.21 tonnes); Upgrading light bulbs (0.10 tonnes); etc. The researchers found that public discourse on reducing one's carbon footprint overwhelmingly focuses on low-impact behaviors, and that mention of the high-impact behaviors is almost non-existent in the mainstream media, government publications, K-12 school textbooks, etc.

The researchers added that "Our recommended high-impact actions are more effective than many more commonly discussed options (e.g. eating a plant-based diet saves eight times more emissions than upgrading light bulbs). More significantly, a US family who chooses to have one fewer child would provide the same level of emissions reductions as 684 teenagers who choose to adopt comprehensive recycling for the rest of their lives."

Dietary Change

Overall, food accounts for the largest share of consumption-based GHG emissions with nearly 20% of the global carbon footprint, followed by housing, mobility, services, manufactured products, and construction. Food and services are more significant in poor countries, while mobility and manufactured goods are more significant in rich countries. A 2014 study into the real-life diets of British people estimates their greenhouse gas contributions (CO_2) to be: 7.19 kg/day for high meat-eaters through to 3.81 kg/day for vegetarians and 2.89 kg/day for vegans. The widespread adoption of a vegetarian diet could cut food-related greenhouse gas emissions by 63% by 2050. China introduced new dietary guidelines in 2016 which aim to cut meat consumption by 50% and thereby reduce greenhouse gas emissions by 1 billion tonnes by 2030. A 2016 study concluded that taxes on meat and milk could simultaneously result in reduced greenhouse gas emissions and healthier diets. The study analyzed surcharges of 40% on beef and 20% on milk and suggests that an optimum plan would reduce emissions by 1 billion tonnes per year.

Energy Efficiency and Conservation

A 230-volt LED light bulb, with an E27 base (10 watts, 806 lumens).

Efficient energy use, sometimes simply called "energy efficiency", is the goal of efforts to reduce the amount of energy required to provide products and services. For example, insulating a home allows a building to use less heating and cooling energy to achieve and maintain a comfortable temperature. Installing LED lighting, fluorescent lighting, or natural skylight windows reduces the amount of energy required to attain the same level of illumination compared to using traditional incandescent light bulbs. Compact fluorescent lamps use only 33% of the energy and may last 6 to 10 times longer than incandescent lights. LED lamps use only about 10% of the energy an incandescent lamp requires.

Energy efficiency has proved to be a cost-effective strategy for building economies without necessarily growing energy consumption. For example, the state of California began implementing energy-efficiency measures in the mid-1970s, including building code and appliance standards with strict efficiency requirements. During the following years, California's energy consumption has remained approximately flat on a per capita basis while national US consumption doubled. As part of its strategy, California implemented a "loading order" for new energy resources that puts energy efficiency first, renewable electricity supplies second, and new fossil-fired power plants last.

Energy conservation is broader than energy efficiency in that it encompasses using less energy to achieve a lesser energy demanding service, for example through behavioral change, as well as encompassing energy efficiency. Examples of conservation without efficiency improvements would be heating a room less in winter, driving less, or working in a less brightly lit room. As with other definitions, the boundary between efficient energy use and energy conservation can be fuzzy, but both are important in environmental and economic terms. This is especially the case when actions are directed at the saving of fossil fuels.

Reducing energy use is seen as a key solution to the problem of reducing greenhouse gas emissions. According to the International Energy Agency, improved energy efficiency in buildings, industrial processes and transportation could reduce the world's energy needs in 2050 by one third, and help control global emissions of greenhouse gases.

Demand-side Switching Sources

Fuel switching on the demand side refers to changing the type of fuel used to satisfy a need for an energy service. To meet deep decarbonization goals, like the 80% reduction by 2050 goal being discussed in California and the European Union, many primary energy changes are needed. Energy efficiency alone may not be sufficient to meet these goals, switching fuels used on the demand side will help lower carbon emissions. Progressively coal, oil and eventually natural gas for space and water heating in buildings will need to be reduced. For an equivalent amount of heat, burning natural gas produces about 45 per cent less carbon dioxide than burning coal. There are various ways in which this could happen, and different strategies will likely make sense in different locations. While the system efficiency of a gas furnace may be higher than the combination of natural gas power plant and electric heat, the combination of the same natural gas power plant and an electric heat pump has lower emissions per unit of heat delivered in all but the coldest climates. This is possible because of the very efficient coefficient of performance of heat pumps.

At the beginning of this century 70% of all electricity was generated by fossil fuels, and as carbon free sources eventually make up half of the generation mix, replacing gas or oil furnaces and water heaters with electric ones will have a climate benefit. In areas like Norway, Brazil, and Quebec that have abundant hydroelectricity, electric heat and hot water are common.

The economics of switching the demand side from fossil fuels to electricity for heating, will depend on the price of fuels vs electricity and the relative prices of the equipment. The EIA Annual Energy Outlook 2014 suggests that domestic gas prices will rise faster than electricity prices which will encourage electrification in the coming decades. Electrifying heating loads may also provide a flexible resource that can participate in demand response. Since thermostatically controlled loads have inherent energy storage, electrification of heating could provide a valuable resource to integrate variable renewable resources into the grid.

Alternatives to electrification, include decarbonizing pipeline gas through power to gas, biogas, or other carbon-neutral fuels. A 2015 study by Energy+Environmental Economics shows that a hybrid approach of decarbonizing pipeline gas, electrification, and energy efficiency can meet carbon reduction goals at a similar cost as only electrification and energy efficiency in Southern California.

Demand Side Grid Management

Expanding intermittent electrical sources such as wind power, creates a growing problem balancing grid fluctuations. Some of the plans include building pumped storage or continental super grids costing billions of dollars. However instead of building for more power, there are a variety of ways to affect the size and timing of electricity demand on the consumer side. Designing for reduced demands on a smaller power grid is more efficient and economic than having extra generation and transmission for intermittentcy, power failures and peak demands. Having these abilities is one of the chief aims of a smart grid.

Time of use metering is a common way to motivate electricity users to reduce their peak load consumption. For instance, running dishwashers and laundry at night after the peak has passed, reduces electricity costs.

Dynamic demand plans have devices passively shut off when stress is sensed on the electrical grid. This method may work very well with thermostats, when power on the grid sags a small amount, a low power temperature setting is automatically selected reducing the load on the grid. For instance millions of refrigerators reduce their consumption when clouds pass over solar installations. Consumers would need to have a smart meter in order for the utility to calculate credits.

Demand response devices could receive all sorts of messages from the grid. The message could be a request to use a low power mode similar to dynamic demand, to shut off entirely during a sudden failure on the grid, or notifications about the current and expected prices for power. This would allow electric cars to recharge at the least expensive rates independent of the time of day. The vehicle-to-grid suggestion would use a car's battery or fuel cell to supply the grid temporarily.

Alternative Energy Sources

Renewable Energy

Renewable energy flows involve natural phenomena such as sunlight, wind, rain, tides, plant growth, and geothermal heat, as the International Energy Agency explains:

> "Renewable energy is derived from natural processes that are replenished constantly. In its various forms, it derives directly from the sun, or from heat generated deep within the earth. Included in the definition is electricity and heat generated from solar, wind, ocean, hydropower, biomass, geothermal resources, and biofuels and hydrogen derived from renewable resources".

The 22,500 MW nameplate capacity Three Gorges Dam in the People's Republic of China, the largest hydroelectric power station in the world.

Climate change concerns and the need to reduce carbon emissions are driving increasing growth in the renewable energy industries. Low-carbon renewable energy replaces conventional fossil fuels in three main areas: power generation, hot water/ space heating, and transport fuels. In 2011, the share of renewables in electricity generation worldwide grew for the fourth year in a row to 20.2%. Based on REN21's 2014 report, renewables contributed 19% to supply global energy consumption. This energy consumption is divided as 9% coming from burning biomass, 4.2% as heat energy (non-biomass), 3.8% hydro electricity and 2% as electricity from wind, solar, geothermal, and biomass thermal power plants.

Renewable energy use has grown much faster than anyone anticipated. The Intergovernmental Panel on Climate Change (IPCC) has said that there are few fundamental technological limits to integrating a portfolio of renewable energy technologies to meet most of total global energy demand. At the national level, at least 30 nations around the world already have renewable energy contributing more than 20% of energy supply.

The 150 MW Andasol solar power station is a commercial parabolic trough solar thermal power plant, located in Spain.

As of 2012, renewable energy accounts for almost half of new electricity capacity installed and costs are continuing to fall. Public policy and political leadership helps to "level the playing field" and drive the wider acceptance of renewable energy technologies. As of 2011, 118 countries have targets for their own renewable energy futures, and have enacted wide-ranging public policies to promote renewables. Leading renewable energy companies include BrightSource Energy, First Solar, Gamesa, GE Energy, Goldwind, Sinovel, Suntech, Trina Solar, Vestas, and Yingli.

The incentive to use 100% renewable energy has been created by global warming and other ecological as well as economic concerns. Mark Z. Jacobson says producing all new energy with wind power, solar power, and hydropower by 2030 is feasible and existing energy supply arrangements could be replaced by 2050. Barriers to implementing the renewable energy plan are seen to be "primarily social and political, not technological or economic". Jacobson says that energy costs with a wind, solar, water system should be similar to today's energy costs. According to a 2011 projection by the (IEA) International Energy Agency, solar power generators may produce most of the world's electricity within 50 years, dramatically reducing harmful greenhouse gas emissions. Critics of the "100% renewable energy" approach include Vaclav Smil and James E. Hansen. Smil and Hansen are concerned about the variable output of solar and wind power, NIMBYism, and a lack of infrastructure.

Economic analysts expect market gains for renewable energy (and efficient energy use) following the 2011 Japanese nuclear accidents. In his 2012 State of the Union address, President Barack Obama restated his commitment to renewable energy and mentioned the long-standing Interior Department commitment to permit 10,000 MW of renewable energy projects on public land in 2012. Globally, there are an estimated 3 million direct jobs in renewable energy industries, with about half of them in the biofuels industry.

Some countries, with favorable geography, geology, and weather well suited to an economical exploitation of renewable energy sources, already get most of their electricity from renewables, including from geothermal energy in Iceland (100 percent), and hydroelectric power in Brazil (85 percent), Austria (62 percent), New Zealand (65 percent), and Sweden (54 percent). Renewable power generators are spread across many countries, with wind power providing a significant share of electricity in some regional areas: for example, 14 percent in the US state of Iowa, 40 percent in the northern German state of Schleswig-Holstein, and 20 percent in Denmark. Solar water heating makes an important and growing contribution in many countries, most notably in China, which now has 70 percent of the global total (180 GWth). Worldwide, total installed solar water heating systems meet a portion of the water heating needs of over 70 million households. The use of biomass for heating continues to grow as well. In Sweden, national use of biomass energy has surpassed that of oil. Direct geothermal heating is also growing rapidly.

Renewable biofuels for transportation, such as ethanol fuel and biodiesel, have contributed to a significant decline in oil consumption in the United States since 2006. The 93 billion liters of biofuels produced worldwide in 2009 displaced the equivalent of an estimated 68 billion liters of gasoline, equal to about 5 percent of world gasoline production. Many different biofuel generations can be distinguished, namely 1st, 2nd, 3rd and 4th generation biofuel. Whereas first generation biofuels competed with food production, the later generations no longer had that problem. Also, as 1st generation biofuels also comprise such fuels as palm oil and soy oil (which are drivers for deforestation in rainforests (Brazil, Indonesia, etc.) that issue is also no longer present in later-generation biofuels.

Some of the world's largest solar power stations: Ivanpah (CSP) and Topaz (PV), both in California.

Nuclear Power

Blue Cherenkov light being produced near the core of the Fission powered Advanced Test Reactor.

Since about 2001 the term "nuclear renaissance" has been used to refer to a possible nuclear power industry revival, driven by rising fossil fuel prices and new concerns about meeting greenhouse gas emission limits. However, in March 2011 the Fukushima nuclear disaster in Japan and associated shutdowns at other nuclear facilities raised questions among some commentators over the future of nuclear power. Platts has reported that "the crisis at Japan's Fukushima nuclear plants has prompted leading energy-consuming countries to review the safety of their existing reactors and cast doubt on the speed and scale of planned expansions around the world".

The World Nuclear Association has reported that nuclear electricity generation in 2012 was at its lowest level since 1999. Several previous international studies and assessments, suggested that as part of the portfolio of other low-carbon energy technologies, nuclear power will continue to play a role in reducing greenhouse gas emissions. Historically, nuclear power usage is estimated to have prevented the atmospheric emission of 64 gigatonnes of CO_2-equivalent as of 2013. Public concerns about nuclear power include the fate of spent nuclear fuel, nuclear accidents, security risks, nuclear proliferation, and a concern that nuclear power plants are very expensive. Of these concerns, nuclear accidents and disposal of long-lived radioactive fuel/"waste" have probably had the greatest public impact worldwide. Although generally unaware of it, both of these glaring public concerns are greatly diminished by present passive safety designs, the experimentally proven, "melt-down proof" EBR-II, future molten salt reactors, and the use of conventional and more advanced fuel/"waste" pyroprocessing, with the latter recycling or reprocessing not presently being commonplace as it is often considered to be cheaper to use a once-through nuclear fuel cycle in many countries, depending on the varying levels of intrinsic value given by a society in reducing the long-lived waste in their country, with France doing a considerable amount of reprocessing when compared to the US.

Nuclear power, with a 10.6% share of world electricity production as of 2013, is second only to hydroelectricity as the largest source of low-carbon power. Over 400 reactors generate electricity in 31 countries.

A Yale University review published in the Journal of Industrial Ecology analyzing CO_2 life cycle assessment(LCA) emissions from nuclear power(light water reactors) determined that: "The collective LCA literature indicates that life cycle GHG emissions from nuclear power are only a fraction of traditional fossil sources and comparable to renewable technologies." While some have raised uncertainty surrounding the future GHG emissions of nuclear power as a result of an extreme potential decline in uranium ore grade without a corresponding increase in the efficiency of enrichment methods. In a scenario analysis of future global nuclear development, as it could be effected by a decreasing global uranium market of average ore grade, the analysis determined that depending on conditions, median life cycle nuclear power GHG emissions could be between 9 and 110 g CO_2-eq/kWh by 2050, with the latter high figure being derived from a "worst-case scenario" that is not "considered very robust" by the authors of the paper, as the "ore grade" in the scenario is lower than the uranium concentration in many lignite coal ashes.

Although this future analyses primarily deals with extrapolations for present Generation II reactor technology, the same paper also summarizes the literature on "FBRs"/Fast Breeder Reactors, of which two are in operation as of 2014 with the newest being the BN-800, for these reactors it states that the "median life cycle GHG emissions are similar to or lower than present light water reactors LWRs and purports to consume little or no uranium ore.

In their 2014 report, the IPCC comparison of energy sources global warming potential per unit of electricity generated, which notably included albedo effects, mirror the median emission value derived from the Warner and Heath Yale meta-analysis for the more common non-breeding light water reactors, a CO_2-equivalent value of 12 g CO_2-eq/kWh, which is the lowest global warming forcing of all baseload power sources, with comparable low carbon power baseload sources, such as hydropower and biomass, producing substantially more global warming forcing 24 and 230 g CO_2-eq/kWh respectively.

The Net Benefits of Low and No-Carbon Electricity Technologies which states, after performing an energy and emissions cost analysis, that "The net benefits of new nuclear, hydro, and natural gas combined cycle plants far outweigh the net benefits of new wind or solar plants", with the most cost effective low carbon power technology being determined to be nuclear power.

During his presidential campaign, Barack Obama stated, "Nuclear power represents more than 70% of our noncarbon generated electricity. It is unlikely that we can meet our aggressive climate goals if we eliminate nuclear power as an option."

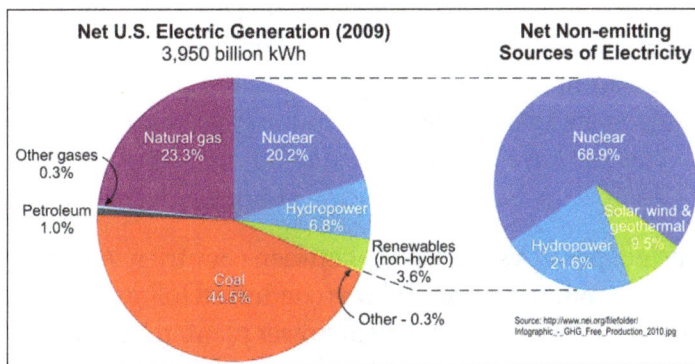

This graph illustrates nuclear power is the United States's largest contributor of non-greenhouse-gas-emitting electric power generation, comprising nearly three-quarters of the non-emitting sources.

Analysis in 2015 by Professor and Chair of Environmental Sustainability Barry W. Brook and his colleagues on the topic of replacing fossil fuels entirely, from the electric grid of the world, has determined that at the historically modest and proven-rate at which nuclear energy was added to and replaced fossil fuels in France and Sweden during each nation's building programs in the 1980s, within 10 years nuclear energy could displace or remove fossil fuels from the electric grid completely, "allowing the world to meet the most stringent greenhouse-gas mitigation targets.". In a similar analysis, Brook had earlier determined that 50% of all global energy, that is not solely electricity, but transportation synfuels etc. could be generated within approximately 30 years, if the global nuclear fission build rate was identical to each of these nation's already proven decadal rates(in units of installed nameplate capacity, GW per year, per unit of global GDP (GW/year/$).

This is in contrast to the completely conceptual paper-studies for a 100% renewable energy world, which would require an orders of magnitude more costly global investment per year, an investment rate that has no historical precedent, having never been attempted due to its prohibitive cost, and with far greater land area that would be required to be devoted to the wind, wave and solar projects, along with the inherent assumption that humanity will use less, and not more, energy in the future. As Brook notes the "principal limitations on nuclear fission are not technical, economic or fuel-related, but are instead linked to complex issues of societal acceptance, fiscal and political inertia, and inadequate critical evaluation of the real-world constraints facing the other low-carbon alternatives."

Nuclear power may be uncompetitive compared with fossil fuel energy sources in countries without a carbon tax program, and in comparison to a fossil fuel plant of the same power output, nuclear power plants take a longer amount of time to construct.

Two new, first of their kind, EPR reactors under construction in Finland and France have been delayed and are running over-budget. However learning from experience, two further EPR reactors under construction in China are on, and ahead, of schedule respectively. As of 2013, according to the IAEA and the European Nuclear Society, worldwide there were 68 civil nuclear power reactors under construction in 15 countries. China has 29 of these nuclear power reactors under construction, as of 2013, with plans to build many more, while in the US the licenses of almost half its reactors have been extended to 60 years, and plans to build another dozen are under serious consideration. There are also a considerable number of new reactors being built in South Korea, India, and Russia. At least 100 older and smaller reactors will "most probably be closed over the next 10–15 years". This is probable only if one does not factor in the ongoing Light Water Reactor Sustainability Program, created to permit the extension of the life span of the USA's 104 nuclear reactors to 60 years. The licenses of almost half of the USA's reactors have been extended to 60 years as of 2008. Two new "passive safety" AP1000 reactors are, as of 2013, being constructed at Vogtle Electric Generating Plant.

Public opinion about nuclear power varies widely between countries. A poll by Gallup International assessed public opinion in 47 countries. The poll was conducted following a tsunami and earthquake which caused an accident at the Fukushima nuclear power plant in Japan. 49% stated that they held favourable views about nuclear energy, while 43% held an unfavourable view. Another global survey by Ipsos assessed public opinion in 24 countries. Respondents to this survey showed a clear preference for renewable energy sources over coal and nuclear energy (refer to graph opposite). Ipsos found that solar and wind were viewed by the public as being more environmentally friendly and more viable long-term energy sources relative to nuclear power and natural gas. However, solar and wind were viewed as being less reliable relative to nuclear power and natural gas. In 2012 a poll done in the UK found that 63% of those surveyed support nuclear power, and with opposition to nuclear power at 11%. In Germany, strong anti-nuclear sentiment led to eight of the seventeen operating reactors being permanently shut down following the March 2011 Fukushima nuclear disaster.

Nuclear fusion research, in the form of the International Thermonuclear Experimental Reactor is underway. Fusion powered electricity generation was initially believed to be readily achievable, as fission power had been. However, the extreme requirements for continuous reactions and plasma containment led to projections being extended by several decades. In 2010, more than 60 years after the first attempts, commercial power production was still believed to be unlikely before 2050. Although rather than an either, or, issue economical fusion-fission hybrid reactors could be built before any attempt at this more demanding commercial "pure-fusion reactor"/DEMO reactor takes place.

Coal to Gas Fuel Switching

Most mitigation proposals imply—rather than directly state—an eventual reduction in global fossil fuel production. Also proposed are direct quotas on global fossil fuel production.

Natural gas emits far fewer greenhouse gases (i.e. CO_2 and methane—CH_4) than coal when burned at power plants, but evidence has been emerging that this benefit could be completely negated by methane leakage at gas drilling fields and other points in the supply chain.

A study performed by the Environmental Protection Agency (EPA) and the Gas Research Institute (GRI) in 1997 sought to discover whether the reduction in carbon dioxide emissions from increased natural gas (predominantly methane) use would be offset by a possible increased level of methane emissions from sources such as leaks and emissions. The study concluded that the reduction in emissions from increased natural gas use outweighs the detrimental effects of increased methane emissions. More recent studies have challenged the findings of this study, with researchers from the National Oceanic and Atmospheric Administration (NOAA) reconfirming findings of high rates of methane (CH_4) leakage from natural gas fields.

A 2011 study by noted climate research scientist, Tom Wigley, found that while carbon dioxide (CO_2) emissions from fossil fuel combustion may be reduced by using natural gas rather than coal to produce energy, it also found that additional methane (CH_4) from leakage adds to the radiative forcing of the climate system, offsetting the reduction in CO_2 forcing that accompanies the transition from coal to gas. The study looked at methane leakage from coal mining; changes in radiative forcing due to changes in the emissions of sulfur dioxide and carbonaceous aerosols; and differences in the efficiency of electricity production between coal- and gas-fired power generation. On balance, these factors more than offset the reduction in warming due to reduced CO_2 emissions. When gas replaces coal there is additional warming out to 2,050 with an assumed leakage rate of 0%, and out to 2,140 if the leakage rate is as high as 10%. The overall effects on global-mean temperature over the 21st century, however, are small. Petron et al. and Alvarez et al. note that estimated that leakage from gas infrastructure is likely to be underestimated. These studies indicate that the exploitation of natural gas as a "cleaner" fuel is questionable. A 2014 meta-study of 20 years of natural gas technical literature shows that methane emissions are consistently underestimated but on a 100-year scale, the climate benefits of coal to gas fuel switching are likely larger than the negative effects of natural gas leakage.

Heat Pump

A heat pump is a device that provides heat energy from a source of heat to a destination called a "heat sink". Heat pumps are designed to move thermal energy opposite to the direction of spontaneous heat flow by absorbing heat from a cold space and releasing it to a warmer one. A heat pump uses some amount of external power to accomplish the work of transferring energy from the heat source to the heat sink.

While air conditioners and freezers are familiar examples of heat pumps, the term "heat pump" is more general and applies to many HVAC (heating, ventilating, and air conditioning) devices used for space heating or space cooling. When a heat pump is used for heating, it employs the same basic refrigeration-type cycle used by an air conditioner or a refrigerator, but in the opposite direction—releasing heat into the conditioned space rather than the surrounding environment. In this use, heat pumps generally draw heat from the cooler external air or from the ground. In heating mode, heat pumps are three to four times more efficient in their use of electric power than simple electrical resistance heaters.

Outside unit of an air-source heat pump.

It has been concluded that heat pumps are the single technology that could reduce the greenhouse gas emissions of households better than every other technology that is available on the market. With a market share of 30% and (potentially) clean electricity, heat pumps could reduce global CO_2 emissions by 8% annually. Using ground source heat pumps could reduce around 60% of the primary energy demand and 90% of CO_2 emissions in Europe in 2050 and make handling high shares of renewable energy easier. Using surplus renewable energy in heat pumps is regarded as the most effective household means to reduce global warming and fossil fuel depletion.

With significant amounts of fossil fuel used in electricity production, demands on the electrical grid also generate greenhouse gases. Without a high share of low-carbon electricity, a domestic heat pump will produce more carbon emissions than using natural gas.

Fossil Fuel Phase-out: Carbon Neutral and Negative Fuels

Fossil fuel may be phased-out with carbon-neutral and carbon-negative pipeline and transportation fuels created with power to gas and gas to liquids technologies. Carbon dioxide from fossil fuel flue gas can be used to produce plastic lumber allowing carbon negative reforestation.

3,500–4,000 environmental activists blocking a coal mine in Germany to limit climate change.

Sinks and Negative Emissions

A carbon sink is a natural or artificial reservoir that accumulates and stores some carbon-containing chemical compound for an indefinite period, such as a growing forest. A negative carbon dioxide emission on the other hand is a permanent removal of carbon dioxide out of the atmosphere. Examples are direct air capture, enhanced weathering technologies such as storing it in geologic formations underground and biochar. These processes are sometimes considered as variations of sinks or mitigation, and sometimes as geoengineering. In combination with other mitigation measures, sinks in combination with negative carbon emissions are considered crucial for meeting the 350 ppm target.

The Antarctic Climate and Ecosystems Cooperative Research Centre (ACE-CRC) notes that one third of humankind's annual emissions of CO_2 are absorbed by the oceans. However, this also leads to ocean acidification, with potentially significant impacts on marine life. Acidification lowers the level of carbonate ions available for calcifying organisms to form their shells. These organisms include plankton species that contribute to the foundation of the Southern Ocean food web. However acidification may impact on a broad range of other physiological and ecological processes, such as fish respiration, larval development and changes in the solubility of both nutrients and toxins. Some plants such as seaweed generate oxygen, are farmed, and can act as a source of (third-generation) biofuel (hereby temporarily sequestering carbon as well).

Reforestation, Avoided Deforestation and Afforestation

Transferring land rights to indigenous inhabitants is argued to efficiently conserve forests.

According to a research by Tom Crowther et al, there is still enough room to plant an additional 1.2 trillion trees. This amount of trees would cancel out the last 10 years of CO_2 emissions and sequester 160 billion tons of carbon.This vision is being executed by the Trillion Tree Campaign. According to research conducted at ETH Zurich, restoring all degraded forests all over the world could capture about 205 billion tons of carbon in total (which is about 2/3rd of all carbon emissions, bringing global warming down to below 2 °C).

Almost 20 percent (8 $GtCO_2$/year) of total greenhouse-gas emissions were from deforestation in 2007. It is estimated that avoided deforestation reduces CO_2 emissions at a rate of 1 tonne of CO_2 per $1–5 in opportunity costs from lost agriculture. Reforestation could save at least another 1

GtCO$_2$/year, at an estimated cost of \$5–15/tCO$_2$. Afforestation is where there was previously no forest – such plantations are estimated to have to be prohibitively massive to be reduce emissions by itself.

Transferring rights over land from public domain to its indigenous inhabitants, who have had a stake for millennia in preserving the forests that they depend on, is argued to be a cost effective strategy to conserve forests. This includes the protection of such rights entitled in existing laws, such as India's Forest Rights Act. The transferring of such rights in China, perhaps the largest land reform in modern times, has been argued to have increased forest cover. Granting title of the land has shown to have two or three times less clearing than even state run parks, notably in the Brazilian Amazon. Excluding humans and even evicting inhabitants from protected areas (called "fortress conservation"), sometimes as a result of lobbying by environmental groups, often lead to more exploitation of the land as the native inhabitants then turn to work for extractive companies to survive.

With increased intensive agriculture and urbanization, there is an increase in the amount of abandoned farmland. By some estimates, for every half a hectare of original old-growth forest cut down, more than 20 hectares of new secondary forests are growing, even though they do not have the same biodiversity as the original forests and original forests store 60% more carbon than these new secondary forests. According to a study in *Science*, promoting regrowth on abandoned farmland could offset years of carbon emissions. Research by the university ETH Zurich estimates that Russia, the United States and Canada have the most land suitable for reforestation.

Avoided Desertification

Managed grazing methods are argued to be able to restore grasslands, thereby significantly decreasing atmospheric CO$_2$ levels.

Restoring grasslands store CO$_2$ from the air into plant material. Grazing livestock, usually not left to wander, would eat the grass and would minimize any grass growth. However, grass left alone would eventually grow to cover its own growing buds, preventing them from photosynthesizing and the dying plant would stay in place. A method proposed to restore grasslands uses fences with many small paddocks and moving herds from one paddock to another after a day a two in order to mimic natural grazers and allowing the grass to grow optimally. Additionally, when part of leaf matter is consumed by a herding animal, a corresponding amount of root matter is sloughed off too as it would not be able to sustain the previous amount of root matter and while most of the lost root matter would rot and enter the atmosphere, part of the carbon is sequestered into the soil. It is estimated that increasing the carbon content of the soils in the world's 3.5 billion hectares of agricultural grassland by 1% would offset nearly 12 years of CO_2 emissions. Allan Savory, as part of holistic management, claims that while large herds are often blamed for desertification, prehistoric lands supported large or larger herds and areas where herds were removed in the United States are still desertifying.

Additionally, the global warming induced thawing of the permafrost, which stores about two times the amount of the carbon currently released in the atmosphere, releases the potent greenhouse gas, methane, in a positive feedback cycle that is feared to lead to a tipping point called runaway climate change. A method proposed to prevent such a scenario is to bring back large herbivores such as seen in Pleistocene Park, where their trampling naturally keep the ground cooler by eliminating shrubs and keeping the ground exposed to the cold air.

Carbon Capture and Storage

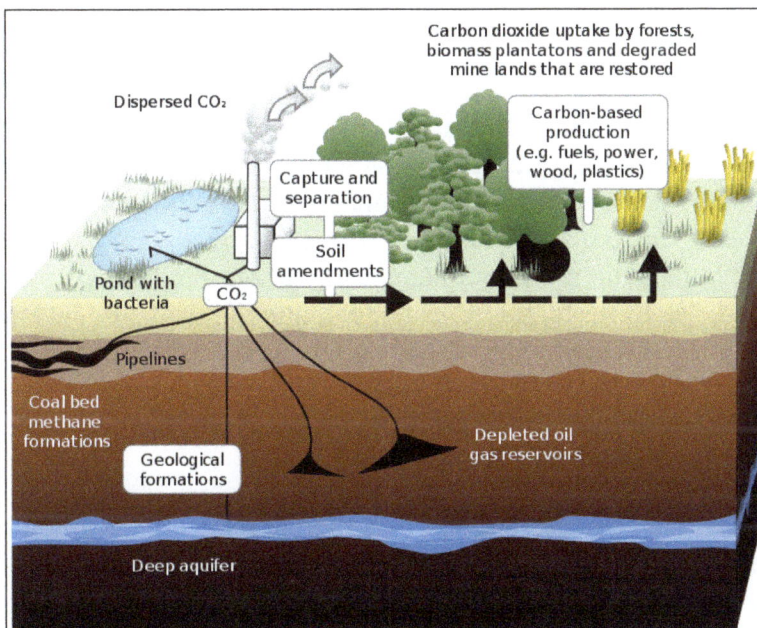

Terrestrial and geological sequestration of carbon dioxide emissions from a coal-fired plant.

Carbon capture and storage (CCS) is a method to mitigate climate change by capturing carbon dioxide (CO_2) from large point sources such as power plants and subsequently storing it away safely instead of releasing it into the atmosphere. The IPCC estimates that the costs of halting global warming would double without CCS. The International Energy Agency says CCS is "the most important single new technology for CO_2 savings" in power generation and industry. Though it requires up to 40% more energy to run a CCS coal power plant than a regular coal plant, CCS could potentially capture about 90% of all the carbon emitted by the plant. Norway's Sleipner gas field, beginning in 1996, stores almost a million tons of CO_2 a year to avoid penalties in producing natural gas with unusually high levels of CO_2. As of late 2011, the total planned CO_2 storage capacity of all 14 projects in operation or under construction is over 33 million tonnes a year. This is broadly equivalent to preventing the emissions from more than six million cars from entering the atmosphere each year. According to a Sierra Club analysis, the

US coal fired Kemper Project due to be online in 2017, is the most expensive power plant ever built for the watts of electricity it will generate.

Enhanced Weathering

Enhanced weathering is the removal of carbon from the air into the earth, enhancing the natural carbon cycle where carbon is mineralized into rock. The CarbFix project couples with carbon capture and storage in power plants to turn carbon dioxide into stone in a relatively short period of two years, addressing the common concern of leakage in CCS projects. While this project used basalt rocks, olivine has also shown promise.

Geoengineering

Geoengineering is seen by Olivier Sterck as an alternative to mitigation and adaptation, but by Gernot Wagner as an entirely separate response to climate change. In a literature assessment, Barker *et al.* described geoengineering as a type of mitigation policy. IPCC concluded that geoengineering options, such as ocean fertilization to remove CO_2 from the atmosphere, remained largely unproven. It was judged that reliable cost estimates for geoengineering had not yet been published.

They evaluated a range of options to try to give preliminary answers to two questions: can these options work and could they be carried out with a reasonable cost. They also sought to encourage discussion of a third question — what adverse side effects might there be. The following types of option were examined: reforestation, increasing ocean absorption of carbon dioxide (carbon sequestration) and screening out some sunlight. NAS also argued "Engineered countermeasures need to be evaluated but should not be implemented without broad understanding of the direct effects and the potential side effects, the ethical issues, and the risks." In July 2011 a report by the United States Government Accountability Office on geoengineering found that "climate engineering technologies do not now offer a viable response to global climate change."

Carbon Dioxide Removal

Carbon dioxide removal has been proposed as a method of reducing the amount of radiative forcing. A variety of means of artificially capturing and storing carbon, as well as of enhancing natural sequestration processes, are being explored. The main natural process is photosynthesis by plants and single-celled organisms. Artificial processes vary, and concerns have been expressed about the long-term effects of some of these processes.

It is notable that the availability of cheap energy and appropriate sites for geological storage of carbon may make carbon dioxide air capture viable commercially. It is, however, generally expected that carbon dioxide air capture may be uneconomic when compared to carbon capture and storage from major sources — in particular, fossil fuel powered power stations, refineries, etc. As in the case of the US Kemper Project with carbon capture, costs of energy produced will grow significantly. However, captured CO_2 can be used to force more crude oil out of oil fields, as Statoil and Shell have made plans to do. CO_2 can also be used in commercial greenhouses, giving an opportunity to kick-start the technology. Some attempts have been made to use algae to capture smokestack emissions, notably the Green Fuel Technologies Corporation, who have now shut down operations.

Solar Radiation Management

The main purpose of solar radiation management seek to reflect sunlight and thus reduce global warming. The ability of stratospheric sulfate aerosols to create a global dimming effect has made them a possible candidate for use in climate engineering projects.

Non-CO_2 Greenhouse Gases

CO_2 is not the only GHG relevant to mitigation, and governments have acted to regulate the emissions of other GHGs emitted by human activities (anthropogenic GHGs). The emissions caps agreed to by most developed countries under the Kyoto Protocol regulate the emissions of almost all the anthropogenic GHGs. These gases are CO_2, methane (CH_4), nitrous oxide (N_2O), the hydrofluorocarbons (HFC), perfluorocarbons (PFC), and sulfur hexafluoride (SF_6).

Stabilizing the atmospheric concentrations of the different anthropogenic GHGs requires an understanding of their different physical properties. Stabilization depends both on how quickly GHGs are added to the atmosphere and how fast they are removed. The rate of removal is measured by the atmospheric lifetime of the GHG in question. Here, the lifetime is defined as the time required for a given perturbation of the GHG in the atmosphere to be reduced to 37% of its initial amount. Methane has a relatively short atmospheric lifetime of about 12 years, while N_2O's lifetime is about 110 years. For methane, a reduction of about 30% below current emission levels would lead to a stabilization in its atmospheric concentration, while for N_2O, an emissions reduction of more than 50% would be required.

Methane is a significantly more potent greenhouse gas than carbon dioxide in the amount of heat it can trap, especially in the short term. Burning one molecule of methane generates one molecule of carbon dioxide, indicating there may be no net benefit in using gas as a fuel source. Reducing the amount of waste methane produced in the first place and moving away from use of gas as a fuel source will have a greater beneficial impact, as might other approaches to productive use of otherwise-wasted methane. In terms of prevention, vaccines are being developed in Australia to reduce the significant global warming contributions from methane released by livestock via flatulence and eructation.

Another physical property of the anthropogenic GHGs relevant to mitigation is the different abilities of the gases to trap heat (in the form of infrared radiation). Some gases are more effective at trapping heat than others, e.g., SF_6 is 22,200 times more effective a GHG than CO_2 on a per-kilogram basis. A measure for this physical property is the global warming potential (GWP), and is used in the Kyoto Protocol.

Although not designed for this purpose, the Montreal Protocol has probably benefited climate change mitigation efforts. The Montreal Protocol is an international treaty that has successfully reduced emissions of ozone-depleting substances (for example, CFCs), which are also greenhouse gases.

By Sector

Transport

Transportation emissions account for roughly 1/4 of emissions worldwide, and are even more important in terms of impact in developed nations especially in North America and Australia. Many citizens of countries like the United States and Canada who drive personal cars often, see well over half of their climate change impact stemming from the emissions produced from their cars. Modes of mass transportation such as bus, light rail (metro, subway, etc.), and long-distance rail are far and away the most energy-efficient means of motorized transportation for passengers, able to use in many cases over twenty times less energy per person-distance than a personal automobile. Modern energy-efficient technologies, such as plug-in hybrid electric vehicles and carbon-neutral synthetic gasoline & Jet fuel may also help to reduce the consumption of petroleum, land use changes and emissions of carbon dioxide. Utilizing rail transport, especially electric rail, over the far less efficient air transport and truck transport significantly reduces emissions. With the use of electric trains and cars in transportation there is the opportunity to run them with low-carbon power, producing far fewer emissions.

The Tesla Roadster emits no tailpipe emissions, uses lithium ion batteries to achieve 220 mi (350 km) per charge, while also capable of going 0–60 in under 4 seconds.

Urban Planning

Bicycles have almost no carbon footprint compared to cars, and canal transport may represent a positive option for certain types of freight in the 21st century.

Effective urban planning to reduce sprawl aims to decrease Vehicle Miles Travelled (VMT), lowering emissions from transportation. Personal cars are extremely inefficient at moving passengers, while public transport and bicycles are many times more efficient (as is the simplest form of human transportation, walking). All of these are encouraged by urban/community planning and are an effective way to reduce greenhouse gas emissions. Between 1982 and 1997, the amount of land consumed for urban development in the United States increased by 47 percent while the nation's population grew by only 17 percent. Inefficient land use development practices have increased infrastructure costs as well as the amount of energy needed for transportation, community services, and buildings.

At the same time, a growing number of citizens and government officials have begun advocating a smarter approach to land use planning. These smart growth practices include compact community development, multiple transportation choices, mixed land uses, and practices to conserve green space. These programs offer environmental, economic, and quality-of-life benefits; and they also serve to reduce energy usage and greenhouse gas emissions.

Approaches such as New Urbanism and transit-oriented development seek to reduce distances travelled, especially by private vehicles, encourage public transit and make walking and cycling more attractive options. This is achieved through "medium-density", mixed-use planning and the concentration of housing within walking distance of town centers and transport nodes.

Smarter growth land use policies have both a direct and indirect effect on energy consuming behavior. For example, transportation energy usage, the number one user of petroleum fuels, could be significantly reduced through more compact and mixed use land development patterns, which in turn could be served by a greater variety of non-automotive based transportation choices.

Building Design

Emissions from housing are substantial, and government-supported energy efficiency programmes can make a difference.

For institutions of higher learning in the United States, greenhouse gas emissions depend primarily on total area of buildings and secondarily on climate. If climate is not taken into account, annual greenhouse gas emissions due to energy consumed on campuses plus purchased electricity can be estimated with the formula, $E = aS^b$, where a =0.001621 metric tonnes of CO_2 equivalent/square foot or 0.0241 metric tonnes of CO_2 equivalent/square meter and b= 1.1354.

New buildings can be constructed using passive solar building design, low-energy building, or zero-energy building techniques, using renewable heat sources. Existing buildings can be made more efficient through the use of insulation, high-efficiency appliances (particularly hot water heaters and furnaces), double- or triple-glazed gas-filled windows, external window shades, and building orientation and siting. Renewable heat sources such as shallow geothermal and passive solar energy reduce the amount of greenhouse gasses emitted. In addition to designing buildings which are more energy-efficient to heat, it is possible to design buildings that are more energy-efficient to cool by using lighter-coloured, more reflective materials in the development of urban areas (e.g. by painting roofs white) and planting trees. This saves energy because it cools buildings and reduces the urban heat island effect thus reducing the use of air conditioning.

Agriculture

In the United States, soils account for about half of agricultural greenhouse gas emissions while agriculture, forestry and other land use emits 24%. Globally, livestock is responsible for 18 percent of greenhouse gas emissions, according to FAO's report called "Livestock's Long Shadow: Environmental Issues and Options".

The US EPA says soil management practices that can reduce the emissions of nitrous oxide (N_2O) from soils include fertilizer usage, irrigation, and tillage. Manure management and rice cultivation also produce gaseous emissions.

Important mitigation options for reducing the greenhouse gas emissions from livestock (especially ruminants) are genetic selection immunization, rumen defaunation, outcompetition of methanogenic archaea with acetogens, introduction of methanotrophic bacteria into the rumen, diet modification and grazing management, among others. Certain diet changes (ie. with Asparagopsis taxiformis) allow for a reduction of upto 99% of ruminant ghg emissions.

Other options include just using ruminant-free alternatives instead, such as milk substitutes and meat analogues.

Non-ruminant livestock (i.e. poultry, etc.) generates far fewer emissions. Methods that enhance carbon sequestration in soil include no-till farming, residue mulching, cover cropping, and crop rotation, all of which are more widely used in organic farming than in conventional farming. Because only 5% of US farmland currently uses no-till and residue mulching, there is a large potential for carbon sequestration.

A 2015 study found that farming can deplete soil carbon and render soil incapable of supporting life; however, the study also showed that conservation farming can protect carbon in soils, and repair damage over time. The farming practise of cover crops has been recognized as climate-smart agriculture by the White House.

In Europe the estimation of the current 0–30 cm SOC stock of agricultural soils was 17.63 Gt. In a subsequent study, authors estimated the best management practices to mitigate soil organic carbon: conversion of arable land to grassland (and vice versa), straw incorporation, reduced tillage, straw incorporation combined with reduced tillage, ley cropping system and cover crops.

An agriculture that mitigates climate change is called Regenerative agriculture. It includes several methods, the main of which are: conservation tillage, diversity, rotation and cover crops, minimizing physical disturbance, minimizing the usage of chemicals. It has other benefits like improving the state of the soil and consequently yields. Some of the big agricultural companies like General Mills and a lot of farms support it.

Societal Controls

Another method being examined is to make carbon a new currency by introducing tradeable "personal carbon credits". The idea being it will encourage and motivate individuals to reduce their 'carbon footprint' by the way they live. Each citizen will receive a free annual quota of carbon that they can use to travel, buy food, and go about their business. It has been suggested that by using this concept it could actually solve two problems; pollution and poverty, old age pensioners will actually be better off because they fly less often, so they can cash in their quota at the end of the year to pay heating bills and so forth.

Population

Various organizations promote population control as a means for mitigating global warming. Proposed measures include improving access to family planning and reproductive health care and information, reducing natalistic politics, public education about the consequences of continued population growth, and improving access of women to education and economic opportunities.

According to a 2017 study published in Environmental Research Letters, having one less child would have a much more substantial effect on greenhouse gas emissions compared with for example living car free or eating a plant-based diet.

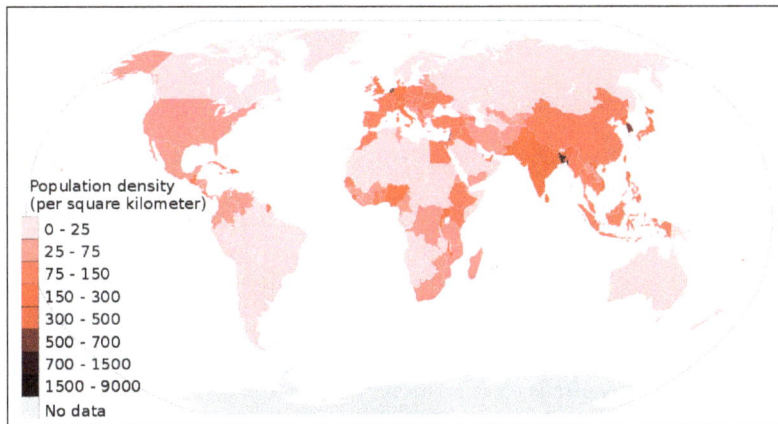

Population density by country.

Population control efforts are impeded by there being somewhat of a taboo in some countries against considering any such efforts. Also, various religions discourage or prohibit some or all forms of birth control.

Population size has a different per capita effect on global warming in different countries, since the per capita production of anthropogenic greenhouse gases varies greatly by country.

Costs and Benefits

Costs

The Stern Review proposes stabilising the concentration of greenhouse-gas emissions in the atmosphere at a maximum of 550ppm CO_2e by 2050. The Review estimates that this would mean cutting total greenhouse-gas emissions to three quarters of 2007 levels. The Review further estimates that the cost of these cuts would be in the range −1.0 to +3.5% of World GDP, (i.e. GWP), with an average estimate of approximately 1%. Stern has since revised his estimate to 2% of GWP. For comparison, the Gross World Product (GWP) at PPP was estimated at $74.5 trillion in 2010, thus 2% is approximately $1.5 trillion. The Review emphasises that these costs are contingent on steady reductions in the cost of low-carbon technologies. Mitigation costs will also vary according to how and when emissions are cut: early, well-planned action will minimise the costs.

One way of estimating the cost of reducing emissions is by considering the likely costs of potential technological and output changes. Policy makers can compare the marginal abatement costs of different methods to assess the cost and amount of possible abatement over time. The marginal abatement costs of the various measures will differ by country, by sector, and over time.

Benefits

Yohe *et al.* assessed the literature on sustainability and climate change. With high confidence, they suggested that up to the year 2050, an effort to cap greenhouse gas (GHG) emissions at 550 ppm would benefit developing countries significantly. This was judged to be especially the case when combined with enhanced adaptation. By 2100, however, it was still judged likely that there would be significant effects of global warming. This was judged to be the case even with aggressive mitigation and significantly enhanced adaptive capacity.

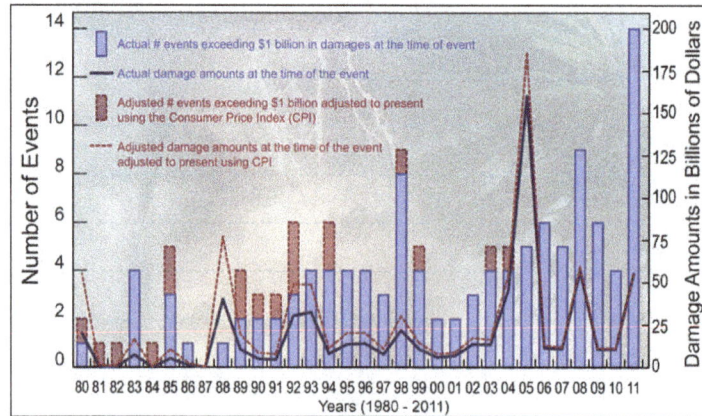

Total extreme weather cost and number of events costing more than $1 billion.

Sharing

One of the aspects of mitigation is how to share the costs and benefits of mitigation policies. There is no scientific consensus over how to share these costs and benefits. In terms of the politics of mitigation, the UNFCCC's ultimate objective is to stabilize concentrations of GHG in the atmosphere at a level that would prevent "dangerous" climate change.

GHG emissions are an important correlate of wealth, at least at present. Wealth, as measured by per capita income (i.e., income per head of population), varies widely between different countries. Activities of the poor that involve emissions of GHGs are often associated with basic needs, such as heating to stay tolerably warm. In richer countries, emissions tend to be associated with things like cars, central heating, etc. The impacts of cutting emissions could therefore have different impacts on human welfare according to wealth.

Distributing Emissions Abatement Costs

There have been different proposals on how to allocate responsibility for cutting emissions:

- Egalitarianism: This system interprets the problem as one where each person has equal rights to a global resource, i.e., polluting the atmosphere.

- Basic needs: This system would have emissions allocated according to basic needs, as defined according to a minimum level of consumption. Consumption above basic needs would require countries to buy more emission rights. From this viewpoint, developing countries would need to be at least as well off under an emissions control regime as they would be outside the regime.

- Proportionality and polluter-pays principle: Proportionality reflects the ancient Aristotelian principle that people should receive in proportion to what they put in, and pay in proportion to the damages they cause. This has a potential relationship with the "polluter-pays principle", which can be interpreted in a number of ways:

  ○ Historical responsibilities: This asserts that allocation of emission rights should be based on patterns of past emissions. Two-thirds of the stock of GHGs in the atmosphere at present is due to the past actions of developed countries.

- ○ Comparable burdens and ability to pay: With this approach, countries would reduce emissions based on comparable burdens and their ability to take on the costs of reduction. Ways to assess burdens include monetary costs per head of population, as well as other, more complex measures, like the UNDP's Human Development Index.

- ○ Willingness to pay: With this approach, countries take on emission reductions based on their ability to pay along with how much they benefit from reducing their emissions.

Specific Proposals

- • Ad hoc: Lashof and Cline, for example, suggested that allocations based partly on GNP could be a way of sharing the burdens of emission reductions. This is because GNP and economic activity are partially tied to carbon emissions.

- • Equal per capita entitlements: This is the most widely cited method of distributing abatement costs, and is derived from egalitarianism. This approach can be divided into two categories. In the first category, emissions are allocated according to national population. In the second category, emissions are allocated in a way that attempts to account for historical (cumulative) emissions.

- • Status quo: With this approach, historical emissions are ignored, and current emission levels are taken as a status quo right to emit. An analogy for this approach can be made with fisheries, which is a common, limited resource. The analogy would be with the atmosphere, which can be viewed as an exhaustible natural resource. In international law, one state recognized the long-established use of another state's use of the fisheries resource. It was also recognized by the state that part of the other state's economy was dependent on that resource.

Governmental and Intergovernmental Action

Many countries, both developing and developed, are aiming to use cleaner technologies. Use of these technologies aids mitigation and could result in substantial reductions in CO_2 emissions. Policies include targets for emissions reductions, increased use of renewable energy, and increased energy efficiency. It is often argued that the results of climate change are more damaging in poor nations, where infrastructures are weak and few social services exist. The Commitment to Development Index is one attempt to analyze rich country policies taken to reduce their disproportionate use of the global commons. Countries do well if their greenhouse gas emissions are falling, if their gas taxes are high, if they do not subsidize the fishing industry, if they have a low fossil fuel rate per capita, and if they control imports of illegally cut tropical timber.

Kyoto Protocol

The main current international agreement on combating climate change is the Kyoto Protocol (which is now succeeded by the Paris agreement). On the 11th of December 1997 the Kyoto Protocol was implemented by the 3rd Conference of Parties, which was coming together in Kyoto, which came into force on 16 February 2005. The Kyoto Protocol is an amendment to the United Nations

Framework Convention on Climate Change (UNFCCC). Countries that have ratified this protocol have committed to reduce their emissions of carbon dioxide and five other greenhouse gases, or engage in emissions trading if they maintain or increase emissions of these gases. For Kyoto reporting, governments are obliged to be told on the present state of the respective countries' forests and the related ongoing processes.

Temperature Targets

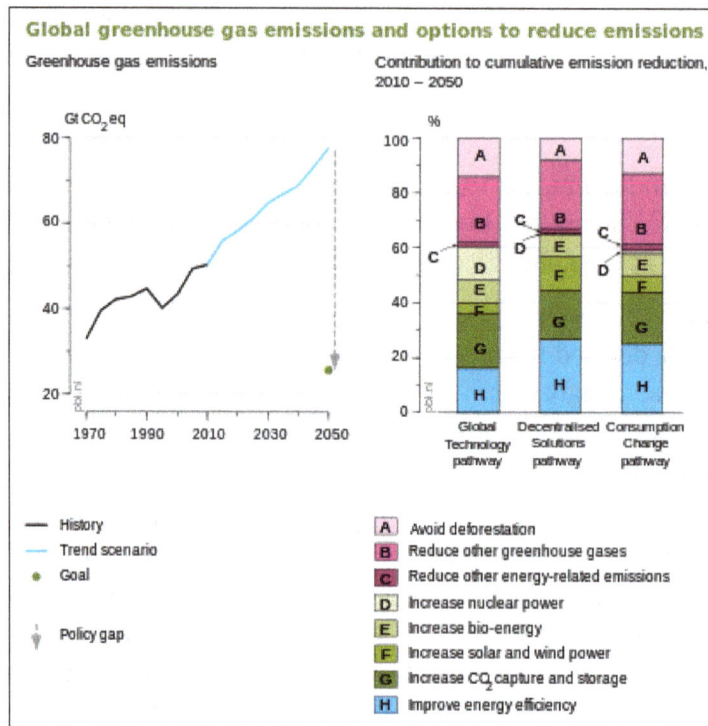

The graph on the right shows three "pathways" to meet the UNFCCC's 2 °C target, labelled "global technology", "decentralised solutions", and "consumption change". Each pathway shows how various measures (e.g., improved energy efficiency, increased use of renewable energy) could contribute to emissions reductions.

Actions to mitigate climate change are sometimes based on the goal of achieving a particular temperature target. One of the targets that has been suggested is to limit the future increase in global mean temperature (global warming) to below 2 °C, relative to the pre-industrial level. The 2 °C target was adopted in 2010 by Parties to the United Nations Framework Convention on Climate Change. Most countries of the world are Parties to the UNFCCC. The target had been adopted in 1996 by the European Union Council.

Feasibility of 2 °C

Temperatures have increased by 0.8 °C compared to the pre-industrial level, and another 0.5–0.7 °C is already committed. The 2 °C rise is typically associated in climate models with a carbon dioxide equivalent concentration of 400–500 ppm by volume; the current (January 2015) level of carbon dioxide alone is 400 ppm by volume, and rising at 1–3 ppm annually. Hence, to avoid a very likely breach of the 2 °C target, CO_2 levels would have to be stabilised very soon; this is generally regarded as unlikely, based on current programs in place to date. The importance of change is illustrated by the fact that world economic energy efficiency is improving at only half the rate of world economic growth.

Views in the Literature

There is disagreement among experts over whether or not the 2 °C target can be met. For example, according to Anderson and Bows, "there is little to no chance" of meeting the target. On the other hand, according to Alcamo:

- Policies adopted by parties to the UNFCCC are too weak to meet a 2 or 1.5 °C target. However, these targets might still be achievable if more stringent mitigation policies are adopted immediately.

- Cost-effective 2 °C scenarios project annual global greenhouse gas emissions to peak before the year 2020, with deep cuts in emissions thereafter, leading to a reduction in 2050 of 41% compared to 1990 levels.

Discussion on other Targets

Scientific analysis can provide information on the impacts of climate change and associated policies, such as reducing GHG emissions. However, deciding what policies are best requires value judgements. For example, limiting global warming to 1 °C relative to pre-industrial levels may help to reduce climate change damages more than a 2 °C limit. However, a 1 °C limit may be more costly to achieve than a 2 °C limit.

According to some analysts, the 2 °C "guardrail" is inadequate for the needed degree and timeliness of mitigation. On the other hand, some economic studies suggest more modest mitigation policies. For example, the emissions reductions proposed by Nordhaus might lead to global warming (in the year 2100) of around 3 °C, relative to pre-industrial levels.

Official Long-term Target of 1.5 °C

In 2015, two official UNFCCC scientific expert bodies came to the conclusion that, "in some regions and vulnerable ecosystems, high risks are projected even for warming above 1.5 °C". This expert position was, together with the strong diplomatic voice of the poorest countries and the island nations in the Pacific, the driving force leading to the decision of the Paris Conference 2015, to lay down this 1.5 °C long-term target on top of the existing 2 °C goal.

Encouraging use Changes

Emissions Tax

Citizens for climate action at the People's Climate March.

An emissions tax on greenhouse gas emissions requires individual emitters to pay a fee, charge or tax for every tonne of greenhouse gas released into the atmosphere. Most environmentally related taxes with implications for greenhouse gas emissions in OECD countries are levied on energy products and motor vehicles, rather than on CO_2 emissions directly. As such, non-transport sectors as the agricultural sector (which too produces potent greenhouse gases, i.e. methane) are typically left untaxed. Also, revenue of the emissions taxes are not always used to offset the emissions directly.

Emission taxes can be both cost-effective and environmentally effective. Difficulties with emission taxes include their potential unpopularity, and the fact that they cannot guarantee a particular level of emissions reduction. Emissions or energy taxes also often fall disproportionately on lower income classes. In developing countries, institutions may be insufficiently developed for the collection of emissions fees from a wide variety of sources.

Subsidies

According to Mark Z. Jacobson, a program of subsidization balanced against expected flood costs could pay for conversion to 100% renewable power by 2030. Jacobson, and his colleague Mark Delucchi, suggest that the cost to generate and transmit power in 2020 will be less than 4 cents per kilowatt hour (in 2007 dollars) for wind, about 4 cents for wave and hydroelectric, from 4 to 7 cents for geothermal, and 8 cents per kWh for solar, fossil, and nuclear power.

Investment

Another indirect method of encouraging uses of renewable energy, and pursue sustainability and environmental protection, is that of prompting investment in this area through legal means, something that is already being done at national level as well as in the field of international investment.

Carbon Emissions Trading

With the creation of a market for trading carbon dioxide emissions within the Kyoto Protocol, it is likely that London financial markets will be the centre for this potentially highly lucrative business; the New York and Chicago stock markets may have a lower trade volume than expected as long as the US maintains its rejection of the Kyoto.

However, emissions trading may delay the phase-out of fossil fuels. In the north-east United States, a successful cap and trade program has shown potential for this solution.

The European Union Emission Trading Scheme (EU ETS) is the largest multi-national, greenhouse gas emissions trading scheme in the world. It commenced operation on 1 January 2005, and all 28 member states of the European Union participate in the scheme which has created a new market in carbon dioxide allowances estimated at 35 billion Euros (US$43 billion) per year. The Chicago Climate Exchange was the first (voluntary) emissions market, and is soon to be followed by Asia's first market (Asia Carbon Exchange). A total of 107 million metric tonnes of carbon dioxide equivalent have been exchanged through projects in 2004, a 38% increase relative to 2003 (78 Mt CO_2e).

Twenty three multinational corporations have come together in the G8 Climate Change Roundtable, a business group formed at the January 2005 World Economic Forum. The group includes

Ford, Toyota, British Airways, and BP. On 9 June 2005 the Group published a statement stating that there was a need to act on climate change and claiming that market-based solutions can help. It called on governments to establish "clear, transparent, and consistent price signals" through "creation of a long-term policy framework" that would include all major producers of greenhouse gases.

The Regional Greenhouse Gas Initiative is a proposed carbon trading scheme being created by nine North-eastern and Mid-Atlantic American states; Connecticut, Delaware, Maine, Massachusetts, New Hampshire, New Jersey, New York, Rhode Island, and Vermont. The scheme was due to be developed by April 2005 but has not yet been completed.

Implementation

Implementation puts into effect climate change mitigation strategies and targets. These can be targets set by international bodies or voluntary action by individuals or institutions. This is the most important, expensive and least appealing aspect of environmental governance.

Funding

Implementation requires funding sources but is often beset by disputes over who should provide funds and under what conditions. A lack of funding can be a barrier to successful strategies as there are no formal arrangements to finance climate change development and implementation. Funding is often provided by nations, groups of nations and increasingly NGO and private sources. These funds are often channelled through the Global Environmental Facility (GEF). This is an environmental funding mechanism in the World Bank which is designed to deal with global environmental issues. The GEF was originally designed to tackle four main areas: biological diversity, climate change, international waters and ozone layer depletion, to which land degradation and persistent organic pollutant were added. The GEF funds projects that are agreed to achieve global environmental benefits that are endorsed by governments and screened by one of the GEF's implementing agencies.

Problems

There are numerous issues which result in a current perceived lack of implementation. It has been suggested that the main barriers to implementation are Uncertainty, Fragmentation, Institutional void, Short time horizon of policies and politicians and Missing motives and willingness to start adapting. The relationships between many climatic processes can cause large levels of uncertainty as they are not fully understood and can be a barrier to implementation. When information on climate change is held between the large numbers of actors involved it can be highly dispersed, context specific or difficult to access causing fragmentation to be a barrier. Institutional void is the lack of commonly accepted rules and norms for policy processes to take place, calling into question the legitimacy and efficacy of policy processes. The Short time horizon of policies and politicians often means that climate change policies are not implemented in favour of socially favoured societal issues. Statements are often posed to keep the illusion of political action to prevent or postpone decisions being made. Missing motives and willingness to start adapting is a large barrier as it prevents any implementation. The issues that arise with a system which involves international government cooperation, such as cap and trade, could potentially be improved with a polycentric approach where the rules are enforced by many small sections of authority as opposed to one overall enforcement agency. Concerns about metal requirement and availability for essential decarbonization technoloqies such as photovoltaics, nuclear power, and (plug-in hybrid) electric vehicles have also been expressed as obstacles.

Occurrence

Despite a perceived lack of occurrence, evidence of implementation is emerging internationally. Some examples of this are the initiation of NAPA's and of joint implementation. Many developing nations have made National Adaptation Programs of Action (NAPAs) which are frameworks to prioritize adaption needs. The implementation of many of these is supported by GEF agencies. Many developed countries are implementing 'first generation' institutional adaption plans particularly at the state and local government scale. There has also been a push towards joint implementation between countries by the UNFCCC as this has been suggested as a cost-effective way for objectives to be achieved.

Territorial Policies

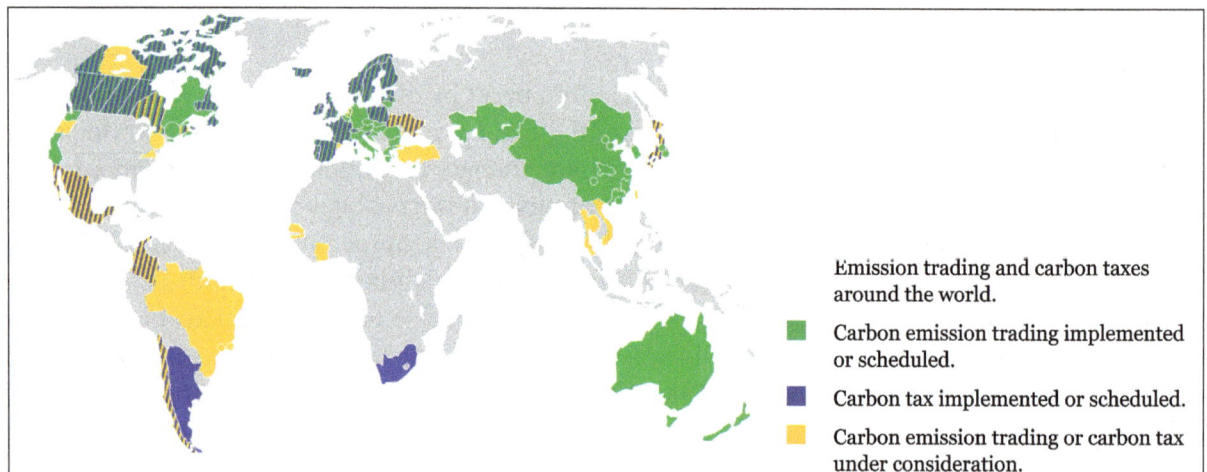

Emission trading and carbon taxes around the world.

■ Carbon emission trading implemented or scheduled.

■ Carbon tax implemented or scheduled.

■ Carbon emission trading or carbon tax under consideration.

North America

Efforts to reduce greenhouse gas emissions by the United States include energy policies which encourage efficiency through programs like Energy Star, Commercial Building Integration, and the Industrial Technologies Program. On 12 November 1998, Vice President Al Gore symbolically signed the Kyoto Protocol, but he indicated participation by the developing nations was necessary prior its being submitted for ratification by the United States Senate.

In 2007, Transportation Secretary Mary Peters, with White House approval, urged governors and dozens of members of the House of Representatives to block California's first-in-the-nation limits on greenhouse gases from cars and trucks, according to e-mails obtained by Congress. The US Climate Change Science Program is a group of about twenty federal agencies and US Cabinet Departments, all working together to address global warming.

The Bush administration pressured American scientists to suppress discussion of global warming, according to the testimony of the Union of Concerned Scientists to the Oversight and Government

Reform Committee of the US House of Representatives. "High-quality science" was "struggling to get out," as the Bush administration pressured scientists to tailor their writings on global warming to fit the Bush administration's skepticism, in some cases at the behest of an ex-oil industry lobbyist. "Nearly half of all respondents perceived or personally experienced pressure to eliminate the words 'climate change,' 'global warming' or other similar terms from a variety of communications." Similarly, according to the testimony of senior officers of the Government Accountability Project, the White House attempted to bury the report "National Assessment of the Potential Consequences of Climate Variability and Change," produced by US scientists pursuant to US law. Some US scientists resigned their jobs rather than give in to White House pressure to underreport global warming.

In the absence of substantial federal action, state governments have adopted emissions-control laws such as the Regional Greenhouse Gas Initiative in the Northeast and the Global Warming Solutions Act of 2006 in California. In 2019 a new climate change bill was introduced in Minnesota. One of the targets, is making all the energy of the state carbon free, by 2030.

European Union

In April 2019 the Spanish Socialist Party takes the biggest number of seats in the elections in Spain. The party declares itself as part of the Green New Deal and propose: "reducing emissions by 90 percent below 1990 levels by mid-century and generating all of the country's power from renewables along the same timeline; 74 percent of power would come from renewables by 2030. The bill would ban fracking nationwide, eliminate fossil fuel subsidies and government investments in fossil fuels, and phase out fossil-fueled vehicles, with the aim of banning the registration and sale of carbon-emitting vehicles by 2040".

Proposals to reach zero carbon economy in the European Union by 2050, were suggested in 2018 – 2019. The countries who support it: Belgium, France, Spain, Netherlands, Denmark, Luxembourg, Portugal, Latvia, Slovenia, Finland, Great Britain, Italy. The main obstacle is Germany, that is not decided yet about its position, and a number of others east – European countries are sceptical also: Bulgaria, Croatia, the Czech Republic and Poland, despite they are the most vulnerable to climate change.

In 2018 was published the Sustainable Action Plan to reorient capital flows toward sustainable investment and manage financial risks from environmental and social problems. In 2019 the European Commission adopted guidelines that help define an environmentally friendly investment. The European Commission is seeking to reduce emissions to zero by 2050. The guidelines set parameters on how businesses can qualify as "green" based on their contributions to the EU's 6 environmental objectives. The guidelines do not include coal and nuclear power and support 2030 goals: at least one-third share of renewable energy in final energy consumption, a one-third energy savings when compared to business-as-usual scenarios, a cut of minimum 40% in greenhouse gas emissions when compared to 1990 levels.

Developing Countries

In order to reconcile economic development with mitigating carbon emissions, developing countries need particular support, both financial and technical. One of the means of achieving this is the Kyoto Protocol's Clean Development Mechanism (CDM). The World Bank's Prototype Carbon Fund is a public private partnership that operates within the CDM.

An important point of contention, however, is how overseas development assistance not directly related to climate change mitigation is affected by funds provided to climate change mitigation. One of the outcomes of the UNFCC Copenhagen Climate Conference was the Copenhagen Accord, in which developed countries promised to provide US$30 million between 2010 and 2012 of new and additional resources. Yet it remains unclear what exactly the definition of additional is and the European Commission has requested its member states to define what they understand to be additional, and researchers at the Overseas Development Institute have found four main understandings:

1. Climate finance classified as aid, but additional to (over and above) the '0.7%' ODA target;

2. Increase on previous year's Official Development Assistance (ODA) spent on climate change mitigation;

3. Rising ODA levels that include climate change finance but where it is limited to a specified percentage; and

4. Increase in climate finance not connected to ODA.

The main point being that there is a conflict between the OECD states budget deficit cuts, the need to help developing countries adapt to develop sustainably and the need to ensure that funding does not come from cutting aid to other important Millennium Development Goals.

However, none of these initiatives suggest a quantitative cap on the emissions from developing countries. This is considered as a particularly difficult policy proposal as the economic growth of developing countries are proportionally reflected in the growth of greenhouse emissions. Critics of mitigation often argue that, the developing countries' drive to attain a comparable living standard to the developed countries would doom the attempt at mitigation of global warming. Critics also argue that holding down emissions would shift the human cost of global warming from a general one to one that was borne most heavily by the poorest populations on the planet.

In an attempt to provide more opportunities for developing countries to adapt clean technologies, UNEP and WTO urged the international community to reduce trade barriers and to conclude the Doha trade round "which includes opening trade in environmental goods and services".

Non-governmental Approaches

While many of the proposed methods of mitigating global warming require governmental funding, legislation and regulatory action, individuals and businesses can also play a part in the mitigation effort.

Choices in Personal Actions and Business Operations

Environmental groups encourage individual action against global warming, often aimed at the consumer. Common recommendations include lowering home heating and cooling usage, burning less gasoline, supporting renewable energy sources, buying local products to reduce transportation, turning off unused devices, and various others.

A geophysicist at Utrecht University has urged similar institutions to hold the vanguard in voluntary mitigation, suggesting the use of communications technologies such as videoconferencing to reduce their dependence on long-haul flights.

Air Travel and Shipment

In 2008, climate scientist Kevin Anderson raised concern about the growing effect of rapidly increasing global air transport on the climate in a paper, and a presentation, suggesting that reversing this trend is necessary to reduce emissions.

Part of the difficulty is that when aviation emissions are made at high altitude, the climate impacts are much greater than otherwise. Others have been raising the related concerns of the increasing hypermobility of individuals, whether traveling for business or pleasure, involving frequent and often long distance air travel, as well as air shipment of goods.

Business Opportunities and Risks

On 9 May 2005 Jeff Immelt, the chief executive of General Electric (GE), announced plans to reduce GE's global warming related emissions by one percent by 2012. "GE said that given its projected growth, those emissions would have risen by 40 percent without such action."

On 21 June 2005 a group of leading airlines, airports, and aerospace manufacturers pledged to work together to reduce the negative environmental impact of aviation, including limiting the impact of air travel on climate change by improving fuel efficiency and reducing carbon dioxide emissions of new aircraft by fifty percent per seat kilometre by 2020 from 2000 levels. The group aims to develop a common reporting system for carbon dioxide emissions per aircraft by the end of 2005, and pressed for the early inclusion of aviation in the European Union's carbon emission trading scheme.

Investor Response

Climate change is also a concern for large institutional investors who have a long term time horizon and potentially large exposure to the negative impacts of global warming because of the large geographic footprint of their multi-national holdings. SRI (Socially responsible investing) Funds allow investors to invest in funds that meet high ESG (environmental, social, governance) standards as such funds invest in companies that are aligned with these goals. Proxy firms can be used to draft guidelines for investment managers that take these concerns into account.

Legal Action

In some countries, those affected by climate change may be able to sue major producers. Attempts at litigation have been initiated by entire peoples such as Palau and the Inuit, as well as non-governmental organizations such as the Sierra Club. Although proving that particular weather events are due specifically to global warming may never be possible, methodologies have been developed to show the increased risk of such events caused by global warming.

For a legal action for negligence (or similar) to succeed, "Plaintiffs must show that, more probably than not, their individual injuries were caused by the risk factor in question, as opposed to any other cause. This has sometimes been translated to a requirement of a relative risk of at least two." Another route (though with little legal bite) is the World Heritage Convention, if it can be shown that climate change is affecting World Heritage Sites like Mount Everest.

Besides countries suing one another, there are also cases where people in a country have taken legal steps against their own government. Legal action for instance has been taken to try to force the US Environmental Protection Agency to regulate greenhouse gas emissions under the Clean Air Act, and against the Export-Import Bank and OPIC for failing to assess environmental impacts (including global warming impacts) under NEPA.

In the Netherlands and Belgium, organisations such as Urgenda and the vzw Klimaatzaak in Belgium have also sued their governments as they believe their governments aren't meeting the emission reductions they agreed to. Urgenda have already won their case against the Dutch government.

According to a 2004 study commissioned by Friends of the Earth, ExxonMobil, and its predecessors caused 4.7 to 5.3 percent of the world's man-made carbon dioxide emissions between 1882 and 2002. The group suggested that such studies could form the basis for eventual legal action.

Exxon received a subpoena. According to the Washington Post and confirmed by the company, the attorney general of New York, Eric Schneiderman, opened an investigation into the possibility that the company had misled the public and investors about the risks of climate change.

22 states, 6 cities and Washington DC in United States, sued the Trump administration for blocking the repealing of Clean power plan.

Activism

Environmental organizations organize different actions such as Peoples Climate Marches and Divestment from fossil fuels. 1,000 organizations with a worth of 8 trillion dollars, made commitments to divest from fossil fuel to 2018. Another form of action is climate strike. In January 2019 12,500 students marched in Brussels demanding Climate action. In 2019 The organization Extinction Rebellion organized massive protests demanding "tell the truth about climate change, reduce carbon emissions to zero by 2025, and create a citizens' assembly to oversee progress", including blocking roads. Many were arrested. In many cases, activism brings positive results.

The main organizators are Fridays For Future and Earth Strike. Trade unions will support the strike The Universities and College Union (UCU) will propose in the next Trades Union Congress (TUC) of England in September to make a workday solidarity stoppage on September 20 in support of the strike. Any walkout would cooperate with the student protest. See more detailes in the page Earth Strike. The target is to influence the climate action summit organized by the UN in 23 of September.

Climate Engineering

Climate engineering is the intentional large-scale intervention in the Earth's climate system to counter climate change. It includes techniques to remove carbon dioxide from the atmosphere, and technologies to rapidly cool the Earth by reflecting solar energy back to space.

Some proposed climate engineering technologies are highly controversial, spurring global debates about whether and under what conditions they should be considered, and reinforcing the pressing need for the US and other nations to commit to aggressive reductions in heat-trapping emissions.

Types of Climate Engineering

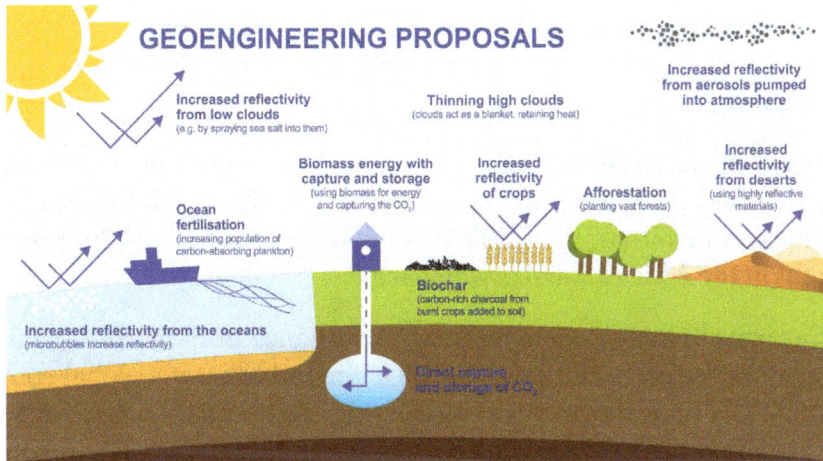

Some carbon dioxide removal techniques (CDR), like reforestation, are well-understood. Others entail using technologies to capture and sequester carbon dioxide that are in early research stages or currently are difficult to deploy at large scales without high costs or substantial negative impacts on energy use, water or land.

Solar geoengineering, or "solar radiation management" (SRM) refers to technologies proposed to rapidly cool down Earth's temperature. Proposals include simulating the cooling effects of volcanic eruptions, and enhancing the reflectivity of marine clouds.

When volcanoes erupt, they spread into the atmosphere tiny particles, commonly known as "aerosols." Light-colored aerosol particles can reflect incoming energy from the sun in cloud-free air and dark particles can absorb it. A small fleet of aircraft, for example, could conceivably inject sulfate aerosols or other reflecting particles into the stratosphere and drive large-scale cooling.

Marine cloud brightening proposals entail using sea salt to "seed"—stimulate the formation of—low-altitude clouds over the ocean, enhance their reflectivity and extend their lifetimes. These SRM techniques are also at early stages of research, primarily based on computer modeling.

SRM technologies would not limit some of the most serious impacts of rising carbon dioxide concentrations, including ocean acidification.

Assessing the Risks and Potential Benefits of Solar Geoengineering

Some climate scientists want to start atmospheric field experiments with sun-reflective aerosols and other solar geoengineering technologies to further understand their risks and potential benefits.

Other scientists and civil society advocates caution, however, that we lack effective mechanisms for global society to consider the risks and potential benefits of these technologies and make informed, well-governed decisions about further research and potential deployment. They warn of the risk, or "moral hazard," that investments in solar geoengineering may diminish efforts at reducing net carbon emissions through proven and affordable means like renewable energy, and that they also may increase geopolitical conflict over "who decides" what the climate goals of deploying SRM would be.

Computer modeling of SRM techniques show, for example, that reflecting sunlight away from the atmosphere may substantially alter rainfall patterns. These preliminary findings indicate that we don't have the full picture of how geoengineering will affect local and regional weather patterns, and therefore geopolitics.

"Power struggles over whose preferences are prioritized in a geo-engineered world appear likely," researchers from UCS and Northeastern University warn. "It is easy to imagine, for example, the US seeking outcomes that maintain favorable precipitation patterns for Midwest farmers even if, in doing so, drought in the African Sahel is worsened."

Forums like the Carnegie Climate Geoengineering Governance Initiative (C_2G_2), and the Solar Radiation Management Governance Initiative (SRMGI) intend to start a global conversation about the scientific and ethical implications of geoengineering. They aim to include different voices and concerns of stakeholders from developed and developing countries about the potential governance of geoengineering research and possible deployment.

With atmospheric field experiments in solar geoengineering proposed to begin in the next few years, there are many issues that should be addressed in advance. These include taking steps to ensure that research is well-governed, that it is carried out only if a broad and representative range of civil society stakeholders grant their informed consent, and that research is funded transparently, and only by governments and other entities fully committed to aggressive reductions in carbon emissions.

Solar geoengineering field research, if it is carried out, must be in the context of a primary commitment from the US and other nations to deep and aggressive reductions in heat-trapping emissions.

Bio-geoengineering

Bio-geoengineering is a form of climate engineering which seeks to use or modify plants or other living things to modify the Earth's climate.

Bio-energy with carbon storage, afforestation projects, and ocean nourishment (including iron fertilization) could be considered examples of bio-geoengineering.

Biogenic aerosols can be grown to replace those beneficial aerosols lost as the result of the death of 50% of Earth's boreal forests. Agricultural production of atmospheric aerosols called "monoterpenes" is possible if crops that are rich in monoterpenes are grown.

Geoengineering is conventionally split into two broad categories. Carbon geoengineering tries to expel carbon dioxide from the environment, which would address the underlying driver of environmental change — the collection of carbon dioxide in the climate. In the chain from outflows to focuses to temperatures to impacts, it breaks the connection from discharges to fixations. Sun powered geoengineering tries to mirror a little part of daylight once more into space or increment the measure of sun oriented radiation that escapes once again into space to cool the planet. Rather than carbon geoengineering, sun based geoengineering does not address the underlying driver of environmental change. It rather expects to break the connection from fixations to temperatures, along these lines diminishing some atmosphere harms. The quick increment in the centralization of air CO_2 proceeded with anthropogenic emanations of this gas is the fundamental factor driving worldwide environmental change. Due to many different causes global temperatures are to increase by 3-5 °C or 5.4 - 9 °F within this century.

Bio-geoengineering Solutions to Climate Change

Carbon Capture and storage: The CO_2 is ordinarily caught before the emissions leave the smokestack, for the most part with a sorbent concoction. The liquified CO_2 is then siphoned into underground aquifers for long haul storage.

Sorbents catch CO_2 from free-streaming air and discharge those atoms as an unadulterated stream of carbon dioxide for sequestration. 100 m² can extract about 1000 tons of carbon dioxide from the atmosphere. 11 million devices would be required to remove 11 billion tons of carbon dioxide.

Stratospheric Aerosols: Amid major volcanic ejections, the Earth regularly experiences huge cooling because of sulfur shot out into the stratosphere. Paul Crutzen a Dutch chemist proposed creating a 'blanket' of sulphur that would block the Sun's rays from reaching Earth.

Seeding the atmosphere to increase reflection of sunlight is also a solution however such approaches are still not being adopted since they have not been proven feasible and have many social, economic, and technical issues before they could be widely adopted.

Downside of Bio-geoengineering

Mirroring the sun's beams into space would modify precipitation designs and reforesting the deserts could change wind designs and could even lessen tree development in different regions. If we are not able to sustain such technology many wildlife can possibly die due to the rapid change in the earth's climate.

Solar Geoengineering

Solar geoengineering, or "solar radiation management" (SRM) refers to technologies proposed to rapidly cool down Earth's temperature.

It is a process through which the reflectivity (albedo) of the Earth's atmosphere or surface is increased, in an attempt to offset some of the effects of GHG-induced climate change. The technique mimics big volcanic eruptions that can cool the Earth by masking the sun with a veil of ash or similar other things.

Methods of Solar Geo Engineering

1. Space-Based Options/Space Sunshades e.g. using mirrors in space, placing vast satellites at Lagrange Point 1, space parasol, etc.

2. Stratosphere-Based Options such as injection of sulphate aerosols into the stratosphere.

3. Cloud-Based Options/Cloud Seeding e.g. Marine Cloud Brightening (by spraying a fine seawater spray in the air), seeding of high cirrus clouds with heterogeneous ice nuclei.

4. Surface-Based Options e.g. whitening roofs, growing more reflective crops, etc.

Solar Radiation Management Governance Initiative

It is an international, NGO-driven project, financed by Dustin Moskovitz (co-founder of Facebook) for expanding the discussion of Solar Radiation Management climate engineering research governance to developing countries. This initiative is a partnership between the Royal Society and the Academy of Sciences for the Developing World (TWAS) and Environmental Defense Fund (EDF).

The prime objective of the initiative to ensure that any geoengineering research that goes ahead – inside or outside the laboratory – is conducted in a manner that is responsible, transparent and environmentally sound.

References

- Climate-change, science: britannica.com, Retrieved 2 March, 2019

- Cronin, thomas n. (2010). Paleoclimates: understanding climate change past and present. New york: columbia university press. Isbn 978-0-231-14494-0

- Global-warming, science: britannica.com, Retrieved 3 May, 2019

- Vaclav smil (2003). The earth's biosphere: evolution, dynamics, and change. Mit press. P. 107. Isbn 978-0-262-69298-4

- Climate-engineering, science, science-and-impacts, global-warming: ucsusa.org, Retrieved 1 August, 2019

- Mcneill, leila. "this lady scientist defined the greenhouse effect but didn't get the credit, because sexism". Smithsonian. Retrieved 2019-04-16

- Climate-impacts-human-health, climate-impacts: epa.gov, Retrieved 11 January, 2019

- Farber, daniel a. (2007). "adapting to climate change: who should pay?". Doi:10.2139/ssrn.980361. Issn 1556-5068

- What-is-climate-change, news, people: lifegate.com, Retrieved 12 July, 2019

- What-is-solar-geoengineering-1548075776-1, general-knowledge: jagranjosh.com, Retrieved 5 March, 2019

Models of Climate

- **General Circulation Model**
- **Earth Systems Model of Intermediate Complexity**

Climate models are systems of differential equations which are used to simulate the interactions of the major drivers of climate such as land surface, oceans, ice and the atmosphere. A few of the important models are atmospheric model and general circulation model. This chapter closely examines these models of climate to provide an extensive understanding of the subject.

Climate models are based on well-documented physical processes to simulate the transfer of energy and materials through the climate system. Climate models, also known as general circulation models or GCMs, use mathematical equations to characterize how energy and matter interact in different parts of the ocean, atmosphere, land. Building and running a climate model is complex process of identifying and quantifying Earth system processes, representing them with mathematical equations, setting variables to represent initial conditions and subsequent changes in climate forcing, and repeatedly solving the equations using powerful supercomputers.

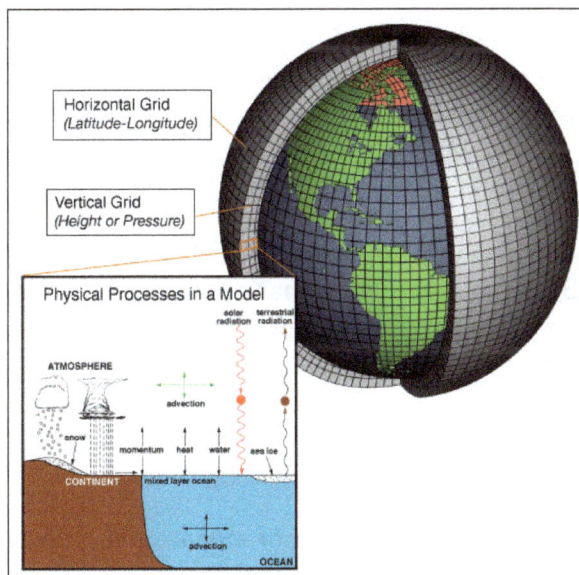

This image shows the concept used in climate models. Each of the thousands of 3-dimensional grid cells can be represented by mathematical equations that describe the materials in it and the way energy moves through it. The advanced equations are based on the fundamental laws of physics, fluid motion, and chemistry. To "run" a model, scientists specify the climate forcing (for instance, setting variables to represent the amount of greenhouse gases in the atmosphere) and have powerful computers solve the equations in each cell. Results from each grid cell are passed to neighboring cells, and the equations are solved again. Repeating the process through many time steps represents the passage of time.

Climate Model Resolution

Climate models separate Earth's surface into a three-dimensional grid of cells. The results of processes modeled in each cell are passed to neighboring cells to model the exchange of matter and energy over time. Grid cell size defines the resolution of the model: the smaller the size of the grid cells, the higher the level of detail in the model. More detailed models have more grid cells, so they need more computing power.

Climate models also include the element of time, called a time step. Time steps can be in minutes, hours, days, or years. Like grid cell size, the smaller the time step, the more detailed the results will be. However, this higher temporal resolution requires additional computing power.

Testing of Climate Models

Once a climate model is set up, it can be tested via a process known as "hind-casting." This process runs the model from the present time backwards into the past. The model results are then compared with observed climate and weather conditions to see how well they match. This testing allows scientists to check the accuracy of the models and, if needed, revise its equations. Science teams around the world test and compare their model outputs to observations and results from other models.

Using Scenarios to Predict Future Climate

Once a climate model can perform well in hind-casting tests, its results for simulating future climate are also assumed to be valid. To project climate into the future, the climate forcing is set to change according to a possible future scenario. Scenarios are possible stories about how quickly human population will grow, how land will be used, how economies will evolve, and the atmospheric conditions (and therefore, climate forcing) that would result for each storyline.

In 2000, the Intergovernmental Panel on Climate Change (IPCC) issued its Special Report on Emissions Scenarios (SRES), describing four scenario families to describe a range of possible future conditions. Referred to by letter-number combinations such as A1, A2, B1, and B2, each scenario was based on a complex relationship between the socioeconomic forces driving greenhouse gas and aerosol emissions and the levels to which those emissions would climb during the 21st century. The SRES scenarios have been in use for more than a decade, so many climate model results describe their inputs using the letter-number combinations.

In 2013, climate scientists agreed upon a new set of scenarios that focused on the level of greenhouse gases in the atmosphere in 2100. Collectively, these scenarios are known as Representative Concentration Pathways or RCPs. Each RCP indicates the amount of climate forcing, expressed in Watts per square meter, that would result from greenhouse gases in the atmosphere in 2100. The rate and trajectory of the forcing is the pathway. Like their predecessors, these values are used in setting up climate models.

Around the world, different teams of scientists have built and run models to project future climate conditions under various scenarios for the next century. The model results project that global temperature will continue to increase, but show that human decisions and behavior we choose today will determine how dramatically climate will change in the future.

Difference between Climate Models and Weather Prediction Models

Unlike weather forecasts, which describe a detailed picture of the expected daily sequence of conditions starting from the present, climate models are probabilistic, indicating areas with higher chances to be warmer or cooler and wetter or drier than usual. Climate models are based on global patterns in the ocean and atmosphere, and records of the types of weather that occurred under similar patterns in the past.

Box Models

Box models are simplified versions of complex systems, reducing them to boxes (or reservoirs) linked by fluxes. The boxes are assumed to be mixed homogeneously. Within a given box, the concentration of any chemical species is therefore uniform. However, the abundance of a species within a given box may vary as a function of time due to the input to (or loss from) the box or due to the production, consumption or decay of this species within the box.

Simple box models, i.e. box model with a small number of boxes whose properties (e.g. their volume) do not change with time, are often useful to derive analytical formulas describing the dynamics and steady-state abundance of a species. More complex box models are usually solved using numerical techniques.

Box models are used extensively to model environmental systems or ecosystems and in studies of ocean circulation and the carbon cycle. They are instances of a multi-compartment model.

Zero-dimensional Models

A very simple model of the radiative equilibrium of the Earth is:

$$(1-a)S\pi r^2 = 4\pi r^2 \epsilon \sigma T^4$$

Where,

- The left hand side represents the incoming energy from the Sun,

- The right hand side represents the outgoing energy from the Earth, calculated from the Stefan-Boltzmann law assuming a model-fictive temperature, T, sometimes called the 'equilibrium temperature of the Earth', that is to be found,

and in the equation above,

- S is the solar constant – the incoming solar radiation per unit area—about 1367 W·m⁻²,

- a is the Earth's average albedo, measured to be 0.3,

- r is Earth's radius—approximately 6.371×10^6 m,

- π is the mathematical constant (3.141...),

- σ is the Stefan-Boltzmann constant—approximately 5.67×10^{-8} J·K^{-4}·m^{-2}·s^{-1},

- $\in$ is the effective emissivity of earth, about 0.612.

The constant πr^2 can be factored out, giving:

$$(1-a)S = 4\epsilon\sigma T^4$$

Solving for the temperature,

$$T = \sqrt[4]{\frac{(1-a)S}{4\epsilon\sigma}}$$

This yields an apparent effective average earth temperature of 288 K (15 °C; 59 °F). This is because the above equation represents the effective *radiative* temperature of the Earth (including the clouds and atmosphere).

This very simple model is quite instructive. For example, it easily determines the effect on average earth temperature of changes in solar constant or change of albedo or effective earth emissivity.

The average emissivity of the earth is readily estimated from available data. The emissivities of terrestrial surfaces are all in the range of 0.96 to 0.99 (except for some small desert areas which may be as low as 0.7). Clouds, however, which cover about half of the earth's surface, have an average emissivity of about 0.5 (which must be reduced by the fourth power of the ratio of cloud absolute temperature to average earth absolute temperature) and an average cloud temperature of about 258 K (–15 °C; 5 °F). Taking all this properly into account results in an effective earth emissivity of about 0.64 (earth average temperature 285 K (12 °C; 53 °F)).

This simple model readily determines the effect of changes in solar output or change of earth albedo or effective earth emissivity on average earth temperature. It says nothing, however about what might cause these things to change. Zero-dimensional models do not address the temperature distribution on the earth or the factors that move energy about the earth.

Radiative-convective Models

The zero-dimensional model above, using the solar constant and given average earth temperature, determines the effective earth emissivity of long wave radiation emitted to space. This can be refined in the vertical to a one-dimensional radiative-convective model, which considers two processes of energy transport:

- Upwelling and downwelling radiative transfer through atmospheric layers that both absorb and emit infrared radiation.

- Upward transport of heat by convection (especially important in the lower troposphere).

The radiative-convective models have advantages over the simple model: they can determine the effects of varying greenhouse gas concentrations on effective emissivity and therefore the surface temperature. But added parameters are needed to determine local emissivity and albedo and address the factors that move energy about the earth.

Effect of ice-albedo feedback on global sensitivity in a one-dimensional radiative-convective climate model.

Higher-dimension Models

The zero-dimensional model may be expanded to consider the energy transported horizontally in the atmosphere. This kind of model may well be zonally averaged. This model has the advantage of allowing a rational dependence of local albedo and emissivity on temperature – the poles can be allowed to be icy and the equator warm – but the lack of true dynamics means that horizontal transports have to be specified.

EMICs (Earth-system Models of Intermediate Complexity)

Depending on the nature of questions asked and the pertinent time scales, there are, on the one extreme, conceptual, more inductive models, and, on the other extreme, general circulation models operating at the highest spatial and temporal resolution currently feasible. Models of intermediate complexity bridge the gap. One example is the Climber-3 model. Its atmosphere is a 2.5-dimensional statistical-dynamical model with $7.5° \times 22.5°$ resolution and time step of half a day; the ocean is MOM-3 (Modular Ocean Model) with a $3.75° \times 3.75°$ grid and 24 vertical levels.

GCMs (global Climate Models or General Circulation Models)

General Circulation Models (GCMs) discretise the equations for fluid motion and energy transfer and integrate these over time. Unlike simpler models, GCMs divide the atmosphere and oceans into grids of discrete "cells", which represent computational units. Unlike simpler models which make mixing assumptions, processes internal to a cell—such as convection—that occur on scales too small to be resolved directly are parameterised at the cell level, while other functions govern the interface between cells.

Atmospheric GCMs (AGCMs) model the atmosphere and impose sea surface temperatures as boundary conditions. Coupled atmosphere-ocean GCMs (AOGCMs, e.g. HadCM3, EdGCM, GFDL CM2.X, ARPEGE-Climat) combine the two models. The first general circulation climate model that combined both oceanic and atmospheric processes was developed in the late 1960s at the NOAA Geophysical Fluid Dynamics Laboratory AOGCMs represent the pinnacle of complexity in climate models and internalise as many processes as possible. However, they are still under development and uncertainties remain. They may be coupled to models of other processes, such as the carbon cycle, so as to better model feedback effects. Such integrated multi-system models are sometimes referred to as either "earth system models" or "global climate models."

Research and Development

There are three major types of institution where climate models are developed, implemented and used:

- National meteorological services. Most national weather services have a climatology section.

- Universities. Relevant departments include atmospheric sciences, meteorology, climatology, and geography.

- National and international research laboratories. Examples include the National Center for Atmospheric Research (NCAR, in Boulder, Colorado, USA), the Geophysical Fluid Dynamics

Laboratory (GFDL, in Princeton, New Jersey, USA), Los Alamos National Laboratory, the Hadley Centre for Climate Prediction and Research (in Exeter, UK), the Max Planck Institute for Meteorology in Hamburg, Germany, or the Laboratoire des Sciences du Climat et de l'Environnement (LSCE), France, to name but a few.

The World Climate Research Programme (WCRP), hosted by the World Meteorological Organization (WMO), coordinates research activities on climate modelling worldwide.

A 2012 U.S. National Research Council report discussed how the large and diverse U.S. climate modeling enterprise could evolve to become more unified. Efficiencies could be gained by developing a common software infrastructure shared by all U.S. climate researchers, and holding an annual climate modeling forum, the report found.

General Circulation Model

A general circulation model (GCM) is a type of climate model. It employs a mathematical model of the general circulation of a planetary atmosphere or ocean. It uses the Navier–Stokes equations on a rotating sphere with thermodynamic terms for various energy sources (radiation, latent heat). These equations are the basis for computer programs used to simulate the Earth's atmosphere or oceans. Atmospheric and oceanic GCMs (AGCM and OGCM) are key components along with sea ice and land-surface components.

GCMs and global climate models are used for weather forecasting, understanding the climate and forecasting climate change.

Versions designed for decade to century time scale climate applications were originally created by Syukuro Manabe and Kirk Bryan at the Geophysical Fluid Dynamics Laboratory (GFDL) in Princeton, New Jersey. These models are based on the integration of a variety of fluid dynamical, chemical and sometimes biological equations.

Atmospheric and Oceanic Models

Atmospheric (AGCMs) and oceanic GCMs (OGCMs) can be coupled to form an atmosphere-ocean coupled general circulation model (CGCM or AOGCM). With the addition of submodels such as a sea ice model or a model for evapotranspiration over land, AOGCMs become the basis for a full climate model.

Trends

A recent trend in GCMs is to apply them as components of Earth system models, e.g. by coupling ice sheet models for the dynamics of the Greenland and Antarctic ice sheets, and one or more chemical transport models (CTMs) for species important to climate. Thus a carbon CTM may allow a GCM to better predict anthropogenic changes in carbon dioxide concentrations. In addition, this approach allows accounting for inter-system feedback: e.g. chemistry-climate models allow the possible effects of climate change on ozone hole to be studied.

Climate prediction uncertainties depend on uncertainties in chemical, physical and social models. Significant uncertainties and unknowns remain, especially regarding the future course of human population, industry and technology.

Structure

Three-dimensional (more properly four-dimensional) GCMs apply discrete equations for fluid motion and integrate these forward in time. They contain parameterisations for processes such as convection that occur on scales too small to be resolved directly.

A simple general circulation model (SGCM) consists of a dynamic core that relates properties such as temperature to others such as pressure and velocity. Examples are programs that solve the primitive equations, given energy input and energy dissipation in the form of scale-dependent friction, so that atmospheric waves with the highest wavenumbers are most attenuated. Such models may be used to study atmospheric processes, but are not suitable for climate projections.

Atmospheric GCMs (AGCMs) model the atmosphere (and typically contain a land-surface model as well) using imposed sea surface temperatures (SSTs). They may include atmospheric chemistry.

AGCMs consist of a dynamical core which integrates the equations of fluid motion, typically for:

- Surface pressure,
- Horizontal components of velocity in layers,
- Temperature and water vapor in layers,
- Radiation, split into solar/short wave and terrestrial/infrared/long wave,
- Parameters for:
  - Convection,
  - Land surface processes,
  - Albedo,
  - Hydrology,
  - Cloud cover.

A GCM contains prognostic equations that are a function of time (typically winds, temperature, moisture, and surface pressure) together with diagnostic equations that are evaluated from them for a specific time period. As an example, pressure at any height can be diagnosed by applying the hydrostatic equation to the predicted surface pressure and the predicted values of temperature between the surface and the height of interest. Pressure is used to compute the pressure gradient force in the time-dependent equation for the winds.

OGCMs model the ocean (with fluxes from the atmosphere imposed) and may contain a sea ice model. For example, the standard resolution of HadOM3 is 1.25 degrees in latitude and longitude, with 20 vertical levels, leading to approximately 1,500,000 variables.

AOGCMs (e.g. HadCM3, GFDL CM2.X) combine the two submodels. They remove the need to specify fluxes across the interface of the ocean surface. These models are the basis for model predictions of future climate, such as are discussed by the IPCC. AOGCMs internalise as many processes as possible. They have been used to provide predictions at a regional scale. While the simpler models are generally susceptible to analysis and their results are easier to understand, AOGCMs may be nearly as hard to analyse as the climate itself.

Grid

The fluid equations for AGCMs are made discrete using either the finite difference method or the spectral method. For finite differences, a grid is imposed on the atmosphere. The simplest grid uses constant angular grid spacing (i.e., a latitude/longitude grid). However, non-rectangular grids (e.g., icosahedral) and grids of variable resolution are more often used. The LMDz model can be arranged to give high resolution over any given section of the planet. HadGEM1 (and other ocean models) use an ocean grid with higher resolution in the tropics to help resolve processes believed to be important for the El Niño Southern Oscillation (ENSO). Spectral models generally use a gaussian grid, because of the mathematics of transformation between spectral and grid-point space. Typical AGCM resolutions are between 1 and 5 degrees in latitude or longitude: HadCM3, for example, uses 3.75 in longitude and 2.5 degrees in latitude, giving a grid of 96 by 73 points (96 x 72 for some variables); and has 19 vertical levels. This results in approximately 500,000 "basic" variables, since each grid point has four variables (u, v, T, Q), though a full count would give more (clouds; soil levels). HadGEM1 uses a grid of 1.875 degrees in longitude and 1.25 in latitude in the atmosphere; HiGEM, a high-resolution variant, uses 1.25 x 0.83 degrees respectively. These resolutions are lower than is typically used for weather forecasting. Ocean resolutions tend to be higher, for example HadCM3 has 6 ocean grid points per atmospheric grid point in the horizontal.

For a standard finite difference model, uniform gridlines converge towards the poles. This would lead to computational instabilities and so the model variables must be filtered along lines of latitude close to the poles. Ocean models suffer from this problem too, unless a rotated grid is used in which the North Pole is shifted onto a nearby landmass. Spectral models do not suffer from this problem. Some experiments use geodesic grids and icosahedral grids, which (being more uniform) do not have pole-problems. Another approach to solving the grid spacing problem is to deform a Cartesian cube such that it covers the surface of a sphere.

Flux Buffering

Some early versions of AOGCMs required an *ad hoc* process of "flux correction" to achieve a stable climate. This resulted from separately prepared ocean and atmospheric models that each used an implicit flux from the other component different than that component could produce. Such a model failed to match observations. However, if the fluxes were 'corrected', the factors that led to these unrealistic fluxes might be unrecognised, which could affect model sensitivity. As a result, the vast majority of models used in the current round of IPCC reports do not use them. The model improvements that now make flux corrections unnecessary include improved ocean physics, improved resolution in both atmosphere and ocean, and more physically consistent coupling between atmosphere and ocean submodels. Improved models now maintain stable, multi-century simulations of surface climate that are considered to be of sufficient quality to allow their use for climate projections.

Convection

Moist convection releases latent heat and is important to the Earth's energy budget. Convection occurs on too small a scale to be resolved by climate models, and hence it must be handled via parameters. This has been done since the 1950s. Akio Arakawa did much of the early work, and variants of his scheme are still used, although a variety of different schemes are now in use. Clouds are also typically handled with a parameter, for a similar lack of scale. Limited understanding of clouds has limited the success of this strategy, but not due to some inherent shortcoming of the method.

Software

Most models include software to diagnose a wide range of variables for comparison with observations or study of atmospheric processes. An example is the 2-metre temperature, which is the standard height for near-surface observations of air temperature. This temperature is not directly predicted from the model but is deduced from surface and lowest-model-layer temperatures. Other software is used for creating plots and animations.

Projections

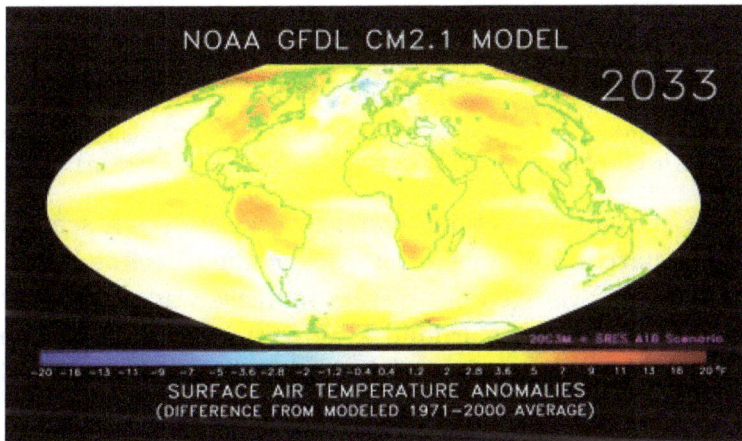

Projected annual mean surface air temperature from 1970-2100, based on SRES emissions scenario A1B, using the NOAA GFDL CM2.1 climate model.

Coupled AOGCMs use transient climate simulations to project/predict climate changes under various scenarios. These can be idealised scenarios (most commonly, CO_2 emissions increasing at 1%/yr) or based on recent history (usually the "IS92a" or more recently the SRES scenarios). Which scenarios are most realistic remains uncertain.

The 2001 IPCC Third Assessment Report figure shows the global mean response of 19 different coupled models to an idealised experiment in which emissions increased at 1% per year. Figure shows the response of a smaller number of models to more recent trends. For the 7 climate models shown there, the temperature change to 2100 varies from 2 to 4.5 °C with a median of about 3 °C.

Future scenarios do not include unknown events – for example, volcanic eruptions or changes in solar forcing. These effects are believed to be small in comparison to greenhouse gas (GHG) forcing in the long term, but large volcanic eruptions, for example, can exert a substantial temporary cooling effect.

Human GHG emissions are a model input, although it is possible to include an economic/technological submodel to provide these as well. Atmospheric GHG levels are usually supplied as an input, though it is possible to include a carbon cycle model that reflects vegetation and oceanic processes to calculate such levels.

Emissions Scenarios

For the six SRES marker scenarios, IPCC gave a "best estimate" of global mean temperature increase (2090–2099 relative to the period 1980–1999) of 1.8 °C to 4.0 °C. Over the same time period, the "likely" range (greater than 66% probability, based on expert judgement) for these scenarios was for a global mean temperature increase of 1.1 to 6.4 °C.

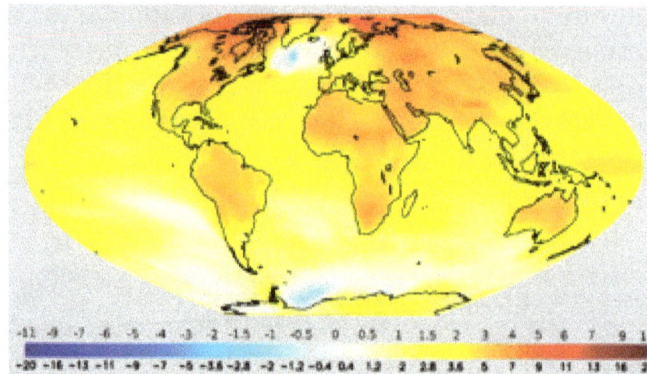

Projected change in annual mean surface air temperature from the late 20th century to the middle 21st century, based on SRES emissions scenario A1B.

In 2008 a study made climate projections using several emission scenarios. In a scenario where global emissions start to decrease by 2010 and then declined at a sustained rate of 3% per year, the likely global average temperature increase was predicted to be 1.7 °C above pre-industrial levels by 2050, rising to around 2 °C by 2100. In a projection designed to simulate a future where no efforts are made to reduce global emissions, the likely rise in global average temperature was predicted to be 5.5 °C by 2100. A rise as high as 7 °C was thought possible, although less likely.

Another no-reduction scenario resulted in a median warming over land (2090–99 relative to the period 1980–99) of 5.1 °C. Under the same emissions scenario but with a different model, the predicted median warming was 4.1 °C.

Model Accuracy

AOGCMs internalise as many processes as are sufficiently understood. However, they are still under development and significant uncertainties remain. They may be coupled to models of other processes, such as the carbon cycle, so as to better model feedbacks. Most recent simulations show "plausible" agreement with the measured temperature anomalies over the past 150 years, when driven by observed changes in greenhouse gases and aerosols. Agreement improves by including both natural and anthropogenic forcings.

Imperfect models may nevertheless produce useful results. GCMs are capable of reproducing the general features of the observed global temperature over the past century.

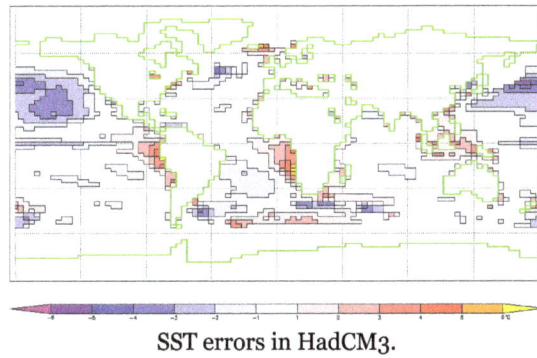

SST errors in HadCM3.

A debate over how to reconcile climate model predictions that upper air (tropospheric) warming should be greater than observed surface warming, some of which appeared to show otherwise, was resolved in favour of the models, following data revisions.

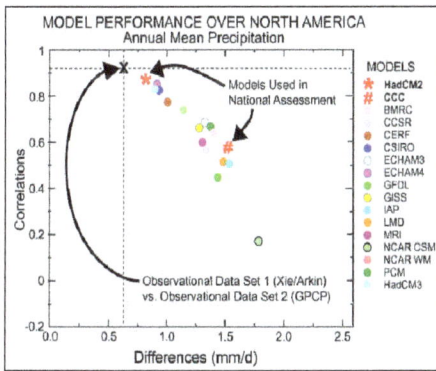

North American precipitation from various models.

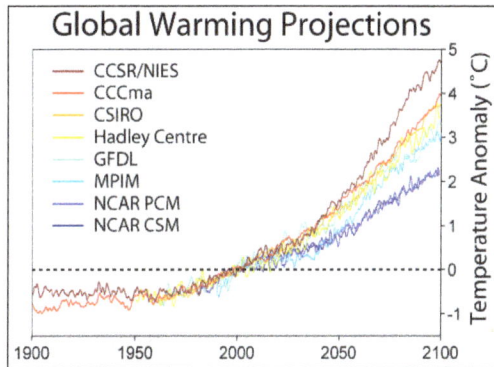

Temperature predictions from some climate models assuming the SRES A2 emissions scenario.

Cloud effects are a significant area of uncertainty in climate models. Clouds have competing effects on climate. They cool the surface by reflecting sunlight into space; they warm it by increasing the amount of infrared radiation transmitted from the atmosphere to the surface. In the 2001 IPCC report possible changes in cloud cover were highlighted as a major uncertainty in predicting climate.

Climate researchers around the world use climate models to understand the climate system. Thousands of papers have been published about model-based studies. Part of this research is to improve the models.

In 2000, a comparison between measurements and dozens of GCM simulations of ENSO-driven tropical precipitation, water vapor, temperature, and outgoing longwave radiation found similarity between measurements and simulation of most factors. However the simulated change in precipitation was about one-fourth less than what was observed. Errors in simulated precipitation imply errors in other processes, such as errors in the evaporation rate that provides moisture to create precipitation. The other possibility is that the satellite-based measurements are in error. Either indicates progress is required in order to monitor and predict such changes.

Climate models is provided in the IPCC's Third Assessment Report:

- The model mean exhibits good agreement with observations.

- The individual models often exhibit worse agreement with observations.

- Many of the non-flux adjusted models suffered from unrealistic climate drift up to about 1 °C/century in global mean surface temperature.

- The errors in model-mean surface air temperature rarely exceed 1 °C over the oceans and 5 °C over the continents; precipitation and sea level pressure errors are relatively greater but the magnitudes and patterns of these quantities are recognisably similar to observations.

- Surface air temperature is particularly well simulated, with nearly all models closely matching the observed magnitude of variance and exhibiting a correlation > 0.95 with the observations.

- Simulated variance of sea level pressure and precipitation is within ±25% of observed.

- All models have shortcomings in their simulations of the present day climate of the stratosphere, which might limit the accuracy of predictions of future climate change.

  ○ There is a tendency for the models to show a global mean cold bias at all levels.

  ○ There is a large scatter in the tropical temperatures.

  ○ The polar night jets in most models are inclined poleward with height, in noticeable contrast to an equatorward inclination of the observed jet.

  ○ There is a differing degree of separation in the models between the winter sub-tropical jet and the polar night jet.

- For nearly all models the r.m.s. error in zonal- and annual-mean surface air temperature is small compared with its natural variability.

  ○ There are problems in simulating natural seasonal variability.

    ▪ In flux-adjusted models, seasonal variations are simulated to within 2 K of observed values over the oceans. The corresponding average over non-flux-adjusted models shows errors up to about 6 K in extensive ocean areas.

    ▪ Near-surface land temperature errors are substantial in the average over flux-adjusted models, which systematically underestimates (by about 5 K) temperature in areas of elevated terrain. The corresponding average over non-flux-adjusted models forms a similar error pattern (with somewhat increased amplitude) over land.

    ▪ In Southern Ocean mid-latitudes, the non-flux-adjusted models overestimate the magnitude of January-minus-July temperature differences by ~5 K due to an overestimate of summer (January) near-surface temperature. This error is common to five of the eight non-flux-adjusted models.

    ▪ Over Northern Hemisphere mid-latitude land areas, zonal mean differences between July and January temperatures simulated by the non-flux-adjusted models show a greater spread (positive and negative) about observed values than results from the flux-adjusted models.

- The ability of coupled GCMs to simulate a reasonable seasonal cycle is a necessary condition for confidence in their prediction of long-term climatic changes (such as global warming), but it is not a sufficient condition unless the seasonal cycle and long-term changes involve similar climatic processes.

- Coupled climate models do not simulate with reasonable accuracy clouds and some related hydrological processes (in particular those involving upper tropospheric humidity). Problems in the simulation of clouds and upper tropospheric humidity, remain worrisome because the associated processes account for most of the uncertainty in climate model simulations of anthropogenic change.

The precise magnitude of future changes in climate is still uncertain; for the end of the 21st century (2071 to 2100), for SRES scenario A2, the change of global average SAT change from AOGCMs compared with 1961 to 1990 is +3.0 °C (5.4 °F) and the range is +1.3 to +4.5 °C (+2.3 to 8.1 °F).

The IPCC's Fifth Assessment Report asserted "very high confidence that models reproduce the general features of the global-scale annual mean surface temperature increase over the historical period". However, the report also observed that the rate of warming over the period 1998–2012 was lower than that predicted by 111 out of 114 Coupled Model Intercomparison Project climate models.

Relation to Weather Forecasting

The global climate models used for climate projections are similar in structure to (and often share computer code with) numerical models for weather prediction, but are nonetheless logically distinct.

Most weather forecasting is done on the basis of interpreting numerical model results. Since forecasts are short – typically a few days or a week – such models do not usually contain an ocean model but rely on imposed SSTs. They also require accurate initial conditions to begin the forecast – typically these are taken from the output of a previous forecast, blended with observations. Predictions must require only a few hours; but because they only cover a one-week the models can be run at higher resolution than in climate mode. Currently the ECMWF runs at 9 km (5.6 mi) resolution as opposed to the 100-to-200 km (62-to-124 mi) scale used by typical climate model runs. Often local models are run using global model results for boundary conditions, to achieve higher local resolution: for example, the Met Office runs a mesoscale model with an 11 km (6.8 mi) resolution covering the UK, and various agencies in the US employ models such as the NGM and NAM models. Like most global numerical weather prediction models such as the GFS, global climate models are often spectral models instead of grid models. Spectral models are often used for global models because some computations in modeling can be performed faster, thus reducing run times.

Computations

Climate models use quantitative methods to simulate the interactions of the atmosphere, oceans, land surface and ice.

All climate models take account of incoming energy as short wave electromagnetic radiation, chiefly visible and short-wave (near) infrared, as well as outgoing energy as long wave (far) infrared electromagnetic radiation from the earth. Any imbalance results in a change in temperature.

The most talked-about models of recent years relate temperature to emissions of greenhouse gases. These models project an upward trend in the surface temperature record, as well as a more rapid increase in temperature at higher altitudes.

Three (or more properly, four since time is also considered) dimensional GCM's discretise the equations for fluid motion and energy transfer and integrate these over time. They also contain parametrisations for processes such as convection that occur on scales too small to be resolved directly.

Atmospheric GCMs (AGCMs) model the atmosphere and impose sea surface temperatures as boundary conditions. Coupled atmosphere-ocean GCMs (AOGCMs, e.g. HadCM3, EdGCM, GFDL CM2.X, ARPEGE-Climat) combine the two models.

Models range in complexity:

- A simple radiant heat transfer model treats the earth as a single point and averages outgoing energy.

- This can be expanded vertically (radiative-convective models), or horizontally.

- Finally, (coupled) atmosphere–ocean–sea ice global climate models discretise and solve the full equations for mass and energy transfer and radiant exchange.

- Box models treat flows across and within ocean basins.

Other submodels can be interlinked, such as land use, allowing researchers to predict the interaction between climate and ecosystems.

Other Climate Models

Earth-system Models of Intermediate Complexity (EMICs)

The Climber-3 model uses a 2.5-dimensional statistical-dynamical model with 7.5° × 22.5° resolution and time step of 1/2 a day. An oceanic submodel is MOM-3 (Modular Ocean Model) with a 3.75° × 3.75° grid and 24 vertical levels.

Radiative-convective Models (RCM)

One-dimensional, radiative-convective models were used to verify basic climate assumptions in the 1980s and 1990s.

Earth Systems Model of Intermediate Complexity

Earth systems Models of Intermediate Complexity (EMICs) form an important class of climate models, primarily used to investigate the earth's systems on long timescales or at reduced computational cost. This is mostly achieved through operation at lower temporal and spatial resolution than more comprehensive general circulation models (GCMs). Due to the non-linear relationship between spatial resolution and model run-speed, modest reductions in resolution can lead to large improvements in model run-speed. This has historically allowed the inclusion of previously unincorporated earth-systems such as ice sheets and carbon cycle feedbacks. These benefits are conventionally understood to come at the cost of some model accuracy. However, the degree to which higher resolution models improve accuracy rather than simply precision is contested.

Computing power had become sufficiently powerful by the middle of the 20th century to allow mass and energy flow models on a vertical and horizontally resolved grid. By 1955 these advances had produced what is recognisable now as a primitive GCM (Phillips prototype). Even at this early stage, a lack of computing power formed a significant barrier to entry and limitation on model-time.

The next half century saw rapid improvement and exponentially increasing computational demands. Modelling on ever smaller length scales required smaller time steps due to the Courant–Friedrichs–Lewy condition. For example, doubling the spatial resolution increases the computational cost by a factor of 16 (factors of 2 for each spatial dimension and time). As well as working on smaller scales, GCMs began to solve more accurate versions of the Navier-Stokes equations. GCMs also began to incorporate more earth systems and feedback mechanisms, transforming themselves into coupled Earth Systems Models. The inclusion of elements from the cryosphere, carbon cycle and cloud feedbacks was both facilitated and constrained by growth in computing power.

The powerful computers and high cost required to run these "comprehensive" models limited accessibility to many university research groups. This helped drive the development of EMICs. Through judicious parametrisation of key variables, researchers could run climate simulations on less powerful computers, or alternatively much faster on comparable computers. A modern example of this difference in speed can be seen between the EMIC JUMP-LCM and the GCM MIROC4h; the former runs 63,000 times faster than the latter. The decrease in required computing power allowed EMICs to run over longer model times, and thus include earth systems occupying the "slow domain".

Petoukhov's 1980 statistical dynamical model has been cited as the first modern EMIC, but despite development throughout the 1980s, their specific value only achieved wider recognition in the late 1990s with inclusion in IPCC AR2 under the moniker of "Simple Climate Models". It was shortly afterwards at the IGBP congress in Shonnan Village, Japan, in May 1999, where the acronym "EMICs" was publicly coined by Claussen. The first simplified model to adopt the nomenclature of "intermediate complexity" is now one of the best known: CLIMBER 2. The Potsdam conference under the guidance of Claussen identified 10 EMICs, a list updated to 13 in 2005. Eight models contributed to IPCC AR4, and 15 to AR4.

Classification of EMICs

As well as "complexity", climate models have been classified by their resolution, parametrisation and "integration". Integration expresses the level of interaction of different components of the earth system. This is influenced by the number of different links in the web (interactivity of coordinates), as well as the frequency of interaction. Because of their speed, EMICs offer the opportunity for highly integrated simulations when compared with more comprehensive ESMs. Four EMIC categorisations have been suggested based on the mode of atmospheric simplification : Statistical-Dynamical Models, Energy Moisture Balance Models, Quasi Geostrophic Models, and

Primitive Equation Models. Of the 15 models in the community contribution to the IPCC's fifth assessment report, four were statistical-dynamic, seven energy moisture balance, two quasi-geostrophic and two primitive equations models.

Statistical Dynamical Models - The CLIMBER Models

CLIMBER-2 and CLIMBER-3α are successive generations of 2.5 and 3 dimensional statistical dynamical models. Rather than continuous evolution of solutions to the Navier Stokes or Primitive Equations, atmospheric dynamics are handled through statistical knowledge of the system (an approach not new to CLIMBER). This approach expresses the dynamics of the atmosphere as large-scale, long term fields of velocity and temperature. Climber-3α's horizontal atmospheric resolution is substantially coarser than a typical atmospheric GCM at 7.5° x 22.5°.

With a characteristic spatial scale of 1000 km, this simplification prohibits resolution of synoptic level features. Climber-3α incorporates comprehensive ocean, sea ice and biogeochemistry models. Despite these full descriptions, simplification of the atmosphere allows it to operate two orders of magnitude faster than comparable GCMs. Both CLIMBER models offer performances comparable to that of contemporary GCMs in simulating present climates. This is clearly of interest due to the significantly lower computational costs. Both models have been principally used to investigate paleoclimates, particularly ice sheet nucleation.

Energy and Moisture Balance Models - UVic ESCM

The thermodynamic approach of the UVic model involves simplification of mass transport (with Fickian diffusion) and precipitation conditions. This model can be seen as a direct descendant of earlier energy balance models. These reductions reduce the atmosphere to three state variables, surface air temperature, sea surface temperature and specific humidity. By parametrising heat and moisture transport with diffusion, timescales are limited to greater than annual and length scales to greater than 1000km. A key result of the thermodynamic rather than fluid dynamic approach is that the simulated climate exhibits no internal variability. Like CLIMBER-3α, it is coupled to a state of the art, 3D ocean model and includes other cutting edge models for sea-ice and land-ice. Unlike CLIMBER, the UVic model does not have significantly coarser resolution than contemporary AOGCMs (3.6° x 1.8°). As such, all computational advantage is from the simplification of atmospheric dynamics.

Quasi Geostrophic Models - LOVECLIM

The quasi-geostrophic equations are a reduction of the Primitive Equations first written down by Charney. These equations are valid in the case of low Rossby number, signifying only a small contribution from inertial forces. Assumed dominance of the Coriolis and pressure-gradient forces facilitates the reduction of the primitive equations to a single equation for potential vorticity in five variables. LOVECLIM features a horizontal resolution of 5.6° and uses the quasi geostrophic atmosphere model ECBilt. It includes a vegetation feedback module by Brovkin et al. The model exhibits some significant limitations that are fundamentally linked to its design. The model predicts an Equilibrium Climate Sensitivity of 1.9°C, at the lower end of the range of GCM predictions. The model's surface temperature distribution is overly-symmetric, and does not represent the northern bias in location of the Intertropical Convergence Zone. The model generally shows lower skill at low latitudes. Other examples of quasi-geostrophic models are PUMA and SPEEDY.

Primitive Equations Model - FAMOUS

The UK Met-Office's FAMOUS blurs the line between more coarsely resolved comprehensive models and EMICs. Designed to run paleoclimate simulations of the Pleistocene, it has been tuned to reproduce the climate of its parent, HADCM3, by solving the Primitive Equations written down by Charney. These are of higher complexity than the quasi-geostrophic equations. Originally named ADTAN, preliminary runs had significant biases involving sea ice and the AMOC, which were later corrected through tuning of sea-ice parameters. The model runs at half the horizontal resolution of HADCM3. Atmospheric resolution is 7.5° x 5°, and oceanic is 3.75° x 2.5°. Atmosphere-Ocean coupling is done once daily.

Comparing and Assessing EMIC Skill

Systematic intercomparison of EMICs has been undertaken since 2000, most recently with a community contribution to the IPCC's fifth assessment report. The equilibrium and transient climate sensitivity of EMICs broadly fell within the range of contemporary GCMs with a range of 1.9 - 4.0°C (compared to 2.1° - 4.7°C, CMIP5). Tested over the last millennium, the average response of the models was close to the real trend, however this conceals much wider variation between individual models. Models generally overestimate ocean heat uptake over the last millennium and indicate a moderate slowing. No relationship was observed in EMICs between levels of polar amplification, climate sensitivity, and initial state. The above comparisons to the performance of GCMs and comprehensive ESMs do not reveal the full value of EMICs. Their ability to run as "fast ESMs" allows them to simulate much longer periods, up to many millennia. As well as running on time-scales far greater than available to GCMs, they provide fertile ground for development and integration of systems that will later join GCMs.

EMICs and the Complexity Spectrum

Possible future directions for EMICs are likely to be in assessment of uncertainties and as a vanguard for incorporation of new earth systems. By virtue of speed they also lend themselves to the creation of ensembles with which to constrain parameters and assess earth systems. EMICs have also recently led in the field of climate stabilisation research. McGuffie and Henderson-Sellers argued in 2001 that in the future, EMICs would be "as important" as GCMs to the climate modelling field - while this has perhaps not been true in the time since that statement, their role has not diminished. Finally, as climate science has come under increasing levels of scrutiny, the ability of models not just to project but to explain has become important. The transparency of EMICs is attractive in this domain, as causal chains are easier to identify and communicate (as opposed to emergent properties generated by comprehensive models).

References

- lynch, peter (2006). "the eniac integrations". The emergence of numerical weather prediction. Cambridge university press. Pp. 206–208. Isbn 978-0-521-85729-1

- Climate-models, primer, maps-data: climate.gov, Retrieved 12 July, 2019

- Marshall, john; plumb, r. Alan (2008). "balanced flow". Atmosphere, ocean, and climate dynamics : an introductory text. Amsterdam: elsevier academic press. Pp. 109–12. Isbn 978-0-12-558691-7

- Mcguffie, k. & a. Henderson-sellers (2005). A climate modelling primer. John wiley and sons. P. 188. Isbn 978-0-470-85751-9

- Sokolov, A.P.; et al. (2009). "Probabilistic Forecast for 21st century Climate Based on Uncertainties in Emissions (without Policy) and Climate Parameters". Journal of Climate. 22 (19): 5175–5204. Bibcode:2009JCLi...22.5175S. doi:10.1175/2009JCLI2863.1

PERMISSIONS

We would like to thank the editorial team for lending their expertise to make the book truly unique. They have played a crucial role in the development of this book. Without their invaluable contributions this book wouldn't have been possible. They have made vital efforts to compile up to date information on the varied aspects of this subject to make this book a valuable addition to the collection of many professionals and students.

This book was conceptualized with the vision of imparting up-to-date and integrated information in this field. To ensure the same, a matchless editorial board was set up. Every individual on the board went through rigorous rounds of assessment to prove their worth. After which they invested a large part of their time researching and compiling the most relevant data for our readers.

The editorial board has been involved in producing this book since its inception. They have spent rigorous hours researching and exploring the diverse topics which have resulted in the successful publishing of this book. They have passed on their knowledge of decades through this book. To expedite this challenging task, the publisher supported the team at every step. A small team of assistant editors was also appointed to further simplify the editing procedure and attain best results for the readers.

Apart from the editorial board, the designing team has also invested a significant amount of their time in understanding the subject and creating the most relevant covers. They scrutinized every image to scout for the most suitable representation of the subject and create an appropriate cover for the book.

The publishing team has been an ardent support to the editorial, designing and production team. Their endless efforts to recruit the best for this project, has resulted in the accomplishment of this book. They are a veteran in the field of academics and their pool of knowledge is as vast as their experience in printing. Their expertise and guidance has proved useful at every step. Their uncompromising quality standards have made this book an exceptional effort. Their encouragement from time to time has been an inspiration for everyone.

The publisher and the editorial board hope that this book will prove to be a valuable piece of knowledge for students, practitioners and scholars across the globe.

INDEX

www.ingramcontent.com/pod-product-compliance
Lightning Source LLC
Chambersburg PA
CBHW061303190326

41458CB00011B/3756